The Human Spermatozoon

Elena Moretti • Giulia Collodel

The Human Spermatozoon

Problems and Approaches

Springer

Elena Moretti
Department of Molecular and Developmental Medicine
University of Siena
Siena, Italy

Giulia Collodel
Department of Molecular and Developmental Medicine
University of Siena
Siena, Italy

ISBN 978-3-031-97867-8 ISBN 978-3-031-97868-5 (eBook)
https://doi.org/10.1007/978-3-031-97868-5

This Springer imprint is published by the registered company Springer Nature Switzerland AG
The registered company address is: Gewerbestrasse 11, 6330 Cham, Switzerland

Preface

Since the discovery of spermatozoa by van Leeuwenhoek in the seventeenth century, it was clear that the spermatozoon is one of the most significant and fascinating cells that has intrigued scientists over time and captured the curiosity of people worldwide. This is likely because the spermatozoon is involved in fertilization, a significant miracle of life linked to the survival of the species.

Over the years, the studies on human spermatozoa have revealed that the inability to conceive is not always the woman's fault, as societal and cultural beliefs have suggested throughout the centuries but can also be linked to the male. This consciousness inevitably influences how male and female roles are perceived in the society.

In clinical practice, much attention is still given to the female aspect by minimizing the male contribution in the search for a reasonably well-shaped and motile spermatozoon assessed under optical microscopy level to be injected into the oocyte. As passionate scholars of spermatozoa for more than 30 years, we thought that presenting a simple book, where this cell was described from different points of view, could support many operators in the field of reproduction, which, unexpectedly, demonstrated a limited knowledge of sperm morphology, particularly from ultrastructural point of view, and of the different alterations affecting sperm functions. Many times, in our professional experiences, we have seen a wrong couple management especially when systematic sperm defects like globozoospermia were present.

Siena (Italy) and its esteemed University, the Master program in "Biotechnologies of Human Reproduction," the course on "Biology of Human Fertility" where Giulia Collodel and I are teaching oogenesis and spermatogenesis: this is the setting that inspired the creation of this book. Nonetheless, the main motivation for writing this book arose from our passionate, curious, and resolute students who possess a deep desire to learn, ask questions, and explore the "human spermatozoon world" further.

This book provides many interesting facts about spermatozoa, and the reader will learn what a beautiful and astonishing cell the spermatozoon is from its generation by spermatogenetic process until its maturation and acquisition of motility during epididymal transit. In addition, the reader will appreciate a technical and simple discussion of alterations and pathologies affecting the spermatogenetic process and a detailed excursus of the ultrastructure of normal and altered human spermatozoon.

We believe that students in Biology, Medicine, Biotechnologies, and Science and all those who work in the field of assisted reproduction as technicians, gynecologists, reproductive biologists, and embryologists can enjoy this work.

Siena, Italy — Elena Moretti
Siena, Italy — Giulia Collodel
May 2025

Acknowledgments

Writing a book is a hard task and we would like to thank several persons that have been part on the realization of this project.

First, we are eternally grateful to Baccio Baccetti, our professor that passed away end of 2010. He transmitted to us the curiosity of studying the spermatozoa from many perspectives. With his huge competences, enthusiasm in approaching sciences, and his ironic attitude, he represented an exceptional guide and a constant support, also now, every day.

We dedicate a warm acknowledgement to Tommaso Renieri, first our professor of Genetics, then our creative and collaborative colleague in our academic journey. We would like to express sincere gratitude and appreciation for his time dedicated, guidance, and positive influence due to his extraordinary humanity, empathy, and supporting ability.

A special mention goes to our smart, professional, and competent friends and colleagues Cinzia Signorini and Cesare Castellini: how many ideas, how much proficient discussions were developed and translated in papers!

We would like to offer our special thanks to our awesome PhD students Laura Liguori and Caterina Marcucci for their support in creating images used in this book. They are always positive and hungry to learn and give us the necessary energy that only young people can radiate. We are deeply grateful also to the designer Leonardo Lazzeri for the priceless advices on the image for the book cover.

We would like to acknowledge Cristina Salvatici for her competence and invaluable assistance in acquiring and discussing transmission electron microscopy micrographs; the time spent at the Centre of Electron Microscopies (CNR, Sesto Fiorentino, Italy) is always an extremely stimulating and productive moment.

We would like to extend our extreme appreciation to the work performed by David Walthall for the careful editing of the entire manuscript and the constructive feedback that was essential for a critical review of this book.

Finally, thanks to everyone in Springer team who helped us so much, and a warm special thanks to Tanja Weyandt, our ever patient Editor, who followed step by step the creation of the book with the right suggestions and advices.

Contents

The History of the Spermatozoon: From Discovery to Today

1

Contents

Core Messages

This chapter traces the history of the acquisition of knowledge about the spermatozoon. From the Greeks to today, writers and thinkers have been fascinated by the mystery of reproduction. Over the centuries, scientists, using the methodologies available to them, have discovered the role and both the normal and pathological structure of sperm. This cell, with its peculiar function, which is to fertilize the oocyte, is different from all the others because it shows motility needed to transfer genetic material to the embryo.

The last part of this chapter addresses the problems associated with the advent of assisted reproduction and the biological perplexities that they have raised. New methodologies are also reported, such as the development of techniques to support in vitro spermatogenesis, the use of artificial intelligence and virtual reality, which could open up new scenarios in the study of reproduction.

E. Moretti, G. Collodel, *The Human Spermatozoon*,
https://doi.org/10.1007/978-3-031-97868-5_1

Abbreviations

AI	Artificial intelligence
ART	Assisted reproduction technologies
CASA	Computer-aided sperm analysis
FISH	Fluorescence in situ hybridization
ICSI	Intracytoplasmic sperm injection
ISS	International Symposium of Comparative Spermatology
IVF	In vitro fertilization
OS	Oxidative stress
SEM	Scanning electron microscopy
TEM	Transmission electron microscopy
VR	Virtual reality
WHO	World Health Organization

1.1 The Origin of the Amazing History of the Spermatozoon

The spermatozoon is one of the most significant and fascinating cells that has intrigued scientists throughout history and captured the curiosity of people worldwide. This is probably because the spermatozoon is involved in fertilization, a meaningful miracle of life linked to the survival of the species.

The aim of this chapter is to outline the key moments in the history of the discovery of spermatozoa morphology and function, mentioning some important figures who made significant contribution in this field of research.

For a complete dissertation, we can refer to some important reviews and book chapters (Afzelius and Baccetti 1991; Cobb 2006; Birkhead and Montgomerie 2009).

At the beginning of this history, a clarification is necessary. "Semen" and "sperm" do not have the same meaning, and this is significant in understanding the different steps of the history of spermatology and the concepts when reading a scientific paper. Semen refers to both seminal fluids produced by accessory glands and spermatozoa produced by the testis; sperm refers to spermatozoa. The semen, the ejaculate, has been known through history, the spermatozoa were observed and described only in 1677.

In antiquity, it was believed that a child was the offspring of either the mother alone or, according to the opposing and rather widespread view, the father alone.

This concept is clearly expressed by Aischylos in the 2500-year-olds play "Eumenides"; here Apollo's speech sounds thus "*The mother is no parent of that which is called her child, but only nurse of the new-planted seed that grows. The parent is the male…*" (Pypłacz 2022). In other words, the seed (male) is sown in a fertile soil (female); it is noteworthy that the word spermatozoon comes from the Greek word "*sperein*" which means "to sow."

Aristotle, whose ideas dominated European thinking for about two millennials, expressed a sophisticated opinion about the role of male and female in the reproductive process: the child is the product of maternal and paternal secretions; however, the semen acts on the female materials as a sculptor acts in a marble block. In other words, the male has a shaping influence on the materials supplied by the female.

During the Renaissance, with a resurgence of interest in the sciences, the researchers were freed from the constriction of Aristotle's ideas, and they were interested in studying reproductive processes.

For instance, Leonardo da Vinci, observing a child born to an Italian woman and an Ethiopian man, deduced that the contributions of female and male were equal. However, during that same period, alchemists attempted to create the *homunculus*, driven by the male-centric fantasy of fathering children without female involvement (Afzelius and Baccetti 1991).

At this point, a figure that is worth mentioning is the Italian anatomist Fabricius (1537–1619), who described the fertilization in *Gallus domesticus* and stated that the vapor of semen is responsible for fertilization; thus, there is no physical contact between semen and eggs.

In 1623, one gets to the heart of the matter with Ludovicus Gardinius (L. du Gardin, 1572–1633) who hypothesized the presence of particles in the semen able to fertilize the eggs.

Obviously, to observe spermatozoa, a microscope is needed; during the late Renaissance Zacharias Janssen (between 1550/1558–1638) invented a sort of microscope using two convex lenses, and from that point the technical equipment progressively improves and with them the scientific findings.

Despite the relative lack of debate surrounding the discovery of spermatozoa (Afzelius and Baccetti 1991), it is historically credited to Anthonie van Leeuwenhoek (1632–1723; Fig. 1.1), an autodidact who dedicated his free time to constructing rudimentary microscopes and employing them to investigate the unknown and mysterious natural world (Birkhead and Montgomerie 2009).

With this new technological tool, in 1677, the enthusiastically curious van Leeuwenhoek observed and described hundreds unknown objects and, among them, also the spermatozoa in the semen of frogs, fishes, and mammals and called them *animalcula*. He concluded that *animalcula* must be a universal feature in male reproduction (Howards 1997). A curiosity is that van Leeuwenhoek in his writing attributed this milestone to "*Dominus Ham,*" that is, a certain Mr. Ham, a student at that time, who never published anything about semen and spermatozoa. Everyone recognizes the importance of Leeuwenhoek as a microscopist, while less known is the inheritance experiment he performed mating a gray rabbit buck with a white rabbit doe. Since the offspring showed a gray color, another evidence of the "seed-soil" opinion was obtained. In any case, at that time, the prevalent opinion was that the embryo grew from the egg alone, after the semen added an "*aura seminalis*" to trigger its development.

The discovery of spermatozoa in the semen inspired Nicolaas Hartsoeker (1656–1725) who published the famous image of the *homunculus* (Fig. 1.2): a spermatozoon containing a little human in his head.

Then, in early 1700 also Dalempatius (François de Plantade) drew homunculi in spermatozoa. The preformation concept—which holds that everything is in the semen but in a miniaturized form and that the development merely consists of

Fig. 1.1 Portrait of Anthonie van Leeuwenhoek, Natural Philosopher and Zoologist (Anthonie van Leeuwenhoek, portrait by Jan Verkolje; in the Rijksmuseum, Amsterdam)

increasing in size—also appeared to be supported by the *humunculus*. The preformationists were both "animalculist" and "ovist." The "ovist" school originated from the book "*De Generatione*" by Harvey in 1651 showing in the frontispiece the mot "*Ex ovo omnia,*", "all (animals) from eggs." Harvey thus believed that the male (semen) played no part in the formation of the fetus. In the same period, another school of thinking, the epigenesis, held that the new individual was a product of both male and female contributions, but as development progressed, new structures emerged. Arguments between preformation and epigenesis have heated until 1800.

During the eighteenth century, the reproductive issues were far from a clarification, even though other milestones were reached. E.F. Geoffroy and C. du Cerf (1704) hypothesized that *animalcula* were necessary for reproduction, even though many great naturalists were against the idea that *animalcula* could play a role in reproduction. For example, Carl Linnaeus (1746) believed that they were inert corpuscles in the semen, other scientists aggregate of molecules or even parasites; the latter being a very popular hypothesis in that period. After many attempts to classify spermatozoa with a genus and specie, in 1827 Carl Ernest von Baer, the first scientist who observed mammalian oocyte, named them with the genus name Spermatozoon, a word that was universally accepted. In 1893, Gabriel Gustav Valentine described the acrosome (named later in 1898 by M. Lenhossék), a sperm-specific organelle widely distributed in the animal kingdom.

Fig. 1.2 Drawing of homunculus contained in the head of a spermatozoon. N. Hartsoeker, Essay de dioptrique reused from the Wellcome Collection https://wellcomecollection.org/works/wsxjcdqe

At that time, the first attempts of artificial insemination played a relevant part in understanding the role of the spermatozoa. The first experiment in Europe was performed by M. Jacobi (1763) in fishes, then Lazzaro Spallanzani (1729–1799), an Italian university-trained and brilliant biologist, developed successful methods for artificial insemination in other animals (Castellani 1973; Sandler 1973). He artificially inseminated a female dog using semen obtained from the reproductive tract of a dead dog, obtaining a live puppy. In addition, Spallanzani observed that sperm cooled in the snow maintained motility, and this represented a sort of first attempt at "sperm cryopreservation" (Puerta Suárez et al. 2018). Despite Spallanzani's significant contributions, the interpretation of his results was insufficient to determine the true role of spermatozoa in fertilization, which was ultimately identified by Prévost and Dumas in 1824.

In the nineteenth century, Rudolf Albert von Koelliker (1817–1905) concluded that semen of all animals contains spermatozoa, which are cells formed in the testes, and that spermatozoa must be in contact with the egg to make fertilization to occur (Adlam 1980; Birkhead and Montgomerie 2009); the head of the spermatozoon is essential for fertilization. Around 1850, Derbès, Dufossé and von Baer documented the spermatozoon-oocyte interaction during fertilization in sea urchin.

Some years later, Oscar Hertwig (1875) realized that fertilization is the combination of sexually differentiated nuclei: "*Fertilization is the union of sexually differentiated nuclei*" is the famous sentence written in his doctoral thesis (Afzelius and Baccetti 1991). He did not, in fact, observe fertilization. It was Hermann Fol (1879) who first saw a spermatozoon fertilize an oocyte.

The twentieth century opened with the outstanding, productive scientist Gustaf Retzius (1842–1919), who described in detail the structure of spermatozoa of more than 400 species, including many rare animals from all six continents. He excelled in technical competence, had the best microscope available in the period, and his drawings of spermatozoa are detailed and beautiful.

His groundbreaking discovery that closely related species have more similar sperm structures than distantly related species made it possible to infer phylogenetic relationships from sperm data. Consequently, in 1991, Afzelius and Baccetti named Retzius the "first comparative spermatologist."

The mid-twentieth century represents a sort of golden age for the study of and advancement in sperm biology. In the middle of this century, many discoveries were made: the sperm-egg interaction in several animal species, the presence of acrosin and hyaluronidase in the acrosome, the acrosomal reaction and the discovery of sperm capacitation by Austin and Chang (Birkhead and Montgomerie 2009).

1.2 The Advent of Electron Microscopy: The Foundation of Comparative Spermatology and the Discovery of Sperm Ultrastructure

Improved microscopes and microscopical techniques were determinant for the further investigation on the spermatozoa. The introduction of transmission electron microscopy (TEM) by Ernst Ruska and Max Knoll in the mid-twentieth century caused a dramatic increase in the knowledge of sperm biology. TEM, scanning electron microscopy (SEM), and freeze-fracture methods allowed the visualization and the study of the sperm subcellular structures. The current knowledge of sperm structure and function benefited greatly from advances in microscopy field and imaging methods.

In that period, the research was focused on the comparative use of sperm ultrastructure to deduce phylogenetic relationship among taxa. Two Swedish biologists—Åke Franzén (1925-still alive) and Bjorn Afzelius (1925–2008)—significantly advanced the study of sperm biology through the investigation of sperm ultrastructure using the newly developed electron microscope. In particular, Åke Franzén studied sperm in about 200 (mostly marine) invertebrate species, linking sperm form and structure to the fertilization environment. The shape of the spermatozoa is thus dependent both on the reproductive biology and on the phylogenetic position. Afzelius provided the first documentation of pores in the sperm carioteca, the first high-magnification image of the acrosome reaction and discovered the dynein arms on the microtubule doublets of the sperm flagellum, thus offering new insight on the mechanism of sperm motility (Rowe and Rosengrave 2020).

These results encouraged studies of the structure and function of sperm as well as the use of sperm ultrastructure as a taxonomic and phylogenetic method and had an impact on the investigations of the role of cilia and flagella in human disease.

To this end, Don Fawcett (1917–2009), in collaboration with Keith Porter (1912–1997), should be acknowledged for their comprehensive comparative anatomical analysis of sperm cell ultrastructure, which included the first description of the ultrastructure of cilia and sperm flagella.

When discussing comparative spermatology, we must also remember the significant contributions of the Italian biologist Baccio Baccetti (1931–2010), who was a delightful mentor and guide.

Baccetti, as a naturalist, studied spermatozoa of many species, in particular in the field of entomology, a theme that brought him to the attention of the international scientific world, rewarding him with honors and glory. Years later, concomitant with his zoological research activity, he focused most of his attention on the study of sperm alterations (Fig. 1.3) that cause infertility in humans interacting with the medical setting, an environment completely different from the naturalistic one.

In 1969, the first International Symposium of Comparative Spermatology (ISS, Fig. 1.4) was held in Rome and Siena organized by Baccio Baccetti; since then, the ISS has been replicated every 4 years, changing its location in Europe, America, Asia, Australia, and Africa.

The themes proposed during ISS have also changed from morphological study of human and animal spermatozoa, functional and molecular research on male

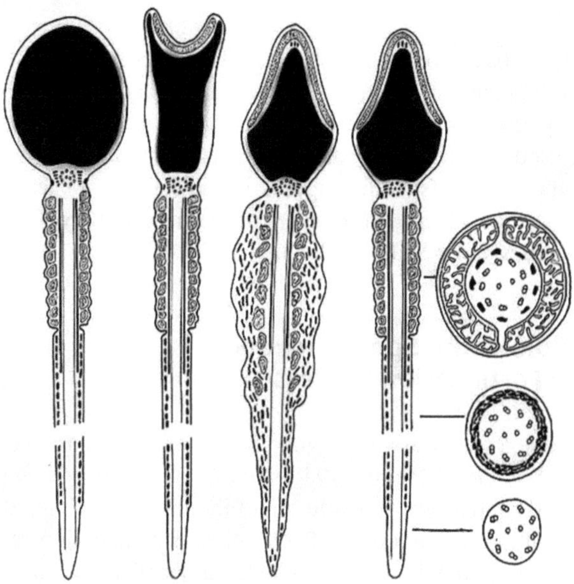

Fig. 1.3 Drawing of longitudinal and cross sections of human spermatozoa observed at transmission electron microscopy level. From left to right: a round-headed spermatozoon without acrosome (globozoospermia), a spermatozoon with an invagination of the nucleus and with an acrosome reduced in dimension (crater defect), a spermatozoon with a short and thick tail with disorganized fibrous sheath (dysplasia of fibrous sheath), a spermatozoon whose axoneme lacks both dynein arms. These defects are classified as systematic or monomorphic sperm alterations. They are rare defects characterized by a predominant anomaly present in most sperm of an ejaculate for the patient's entire life

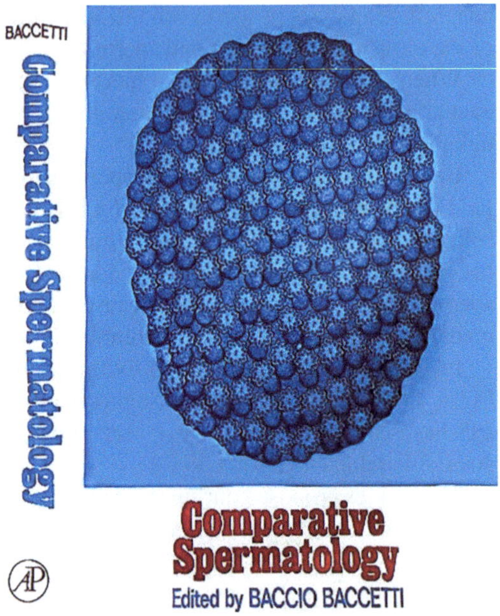

Fig. 1.4 Picture of the book cover of proceeding of the first International Comparative Spermatology meeting held in Rome in 1969

gametes, and from basic to applied spermatology. ISS followed step by step the remarkable development of the studies on spermatozoa and in reproductive biology.

As regards sperm alterations, in particular detected and studied by TEM, it is necessary to introduce Héctor Edgardo Chemes from Buenos Aires, who represented one of the most outstanding figures in the research of testicular and male gamete physiology, introducing the concept of sperm pathology (Chemes and Rawe 2003; Chemes and Alvarez Sedo 2012; Chemes 2018).

1.3 The Role of the Spermatozoon in the Fertilization Process

In the same period, a new second intriguing field of research around in sperm biology was inspired by Geoff Parker (1944-still alive), a world-renowned behavioral ecologist. He instigated a sort of revolution of thought by studying the sperm competition (Birkhead and Montgomerie 2009). Sperm competition refers to the fact that, in many species both with internal or external fertilization, spermatozoa of two or more males compete for egg fertilization. Sexual selection does not stop at mating but continues right up to fertilization, with the obvious consequence that the female of most species are not sexually monogamous, as the Darwinian vision would have wanted (Birkhead 2000). Males, rather than competing for females, compete for fertilization!

Among many outstanding scientists operating in the second half of twentieth century, a special mention goes to Ryuzo Yanagimachi (1928–2023), a

Japanese-born and American-based researcher at University of Hawai'i at Mānoa. He is famous for the cloning of the world's first mouse to survive to adulthood. In 1976, his group injected hamster and human sperm into hamster oocytes and demonstrated that sperm from both species were able to form pronuclei (Uehara and Yanagimachi 1976). Years later, he developed an effective strategy to inject mouse sperm into mouse oocytes. When such in vitro-produced embryos were transferred to surrogate females, they developed into normal pups (Kimura and Yanagimachi 1995). This technique of intracytoplasmic sperm injection (ICSI) was adopted by clinical investigators from Belgium and led to the birth of healthy human babies (Palermo et al. 1992).

In the field of spermatology, Yanagimachi studied the interaction between sperm and zona pellucida of the oocyte and, in particular, in 1981 observed that spermatozoa adopted an extremely vigorous swimming pattern just before the onset of the acrosomal reaction. In other words, he discovered the process of hyperactivation (Yanagimachi 1981).

Through these studies on animal sperm, it was evident that despite spermatozoa shared the same function of fertilizing eggs; they exhibit a tremendous diversity in sperm morphology. It was hypothesized that the environment in which the spermatozoa operate and where the fertilization occurs, the evolutionary history of species and the post-copulatory sexually selective forces of sperm competition are thought to have an impact on morphology. This is an extremely timely topic of research and Fitzpatrick et al. (2022) described a SpermTree database, which represents 5675 morphological descriptions of sperm morphology from 4705 unique species from 27 animal phyla.

Another important sperm parameter is the motility. Around the 1940s and decades later, some scientists realized that they needed objective data on the percentage of motile sperm and, ideally, velocity of the sperm movement. They were motivated by the desire to establish standards and to make subjective evaluations both in animals and human field. The examination of sperm movement characteristics was made possible by the development of computer-aided sperm analysis (CASA) technology in the late 1980s. This field of research has been greatly facilitated by the technology. CASA has been widely utilized in domesticated animal production facilities and also in human research to measure the semen parameters as sperm concentration and progressive motility (Mortimer et al. 2015). Today, a number of different and sophisticated CASA systems are available for the objective evaluation of sperm kinematic parameters, even though highly trained users are required.

In the last decades, the study of sperm biology has been crucial for the field of assisted reproduction technologies (ART) used in the treatment of human infertility but even more in animal breeding contributing to the genetic improvement and supporting the preservation of endangered species and the conservation of biodiversity.

As specifically regards humans, the study of spermatology opened new opportunities to understanding the role of the male in the fertilization process, neglected, or underestimated since antiquity, with cultural implications as well. Indeed, it soon became clear that humans, in the animal kingdom, are peculiar as regards sperm morphology because even healthy and fertile individuals can have altered and poor motile sperm in high percentages. In any case, studies on human spermatozoa have revealed that the responsibility for inability to conceive is not always attributable to the woman, as societal and cultural attitudes seemed imposed over the centuries, but also to the male. These discoveries, together with the increase in the number of couples with fertility problems (Agarwal et al. 2015) and the decline in the semen quality in recent decades (Levine et al. 2023) have triggered a change in the perception of roles within the couple, although still the commonplace belief that infertility depends mostly on female issues has not totally disappeared.

1.4 The Spermatozoa and In Vitro Fertilization

The real breakthrough in the field of human reproduction was reached by Steptoe and Edwards, who published a letter to the Editor in The Lancet in which they reported the first live birth of a "test tube baby": Louise Brown, England 1978. This success was obtained by in vitro fertilization (IVF), recovering oocytes form the mother, fertilizing them under laboratory conditions and finally implanting the obtained embryo. This technique represented an outstanding advance in medical treatment for infertility, and nowadays IVF allows millions of births worldwide. However, it was soon realized that IVF had limitations and couldn't treat all kind of infertility, for example, in the case of severely compromised semen quality, because even though with this technique the gametes are brought close together, the spermatozoon still needs to move, reach the oocyte, and fertilize it.

An exceptional advancement in ART field dates back to 1992 when G. Palermo directly injected a single human spermatozoon in an oocyte (Palermo et al. 1992, 2012) discovering a new technique: ICSI (Fig. 1.5).

Since then, it has been evident that the role of the spermatozoon has completely changed because with ICSI, even an altered immotile spermatozoon can fertilize the oocyte. Therefore, this technique is the most effective one to resolve male gamete dysfunction and affords the best results in terms of fertilization and clinical pregnancy outcome (Palermo et al. 2012). The ICSI procedure significantly altered the

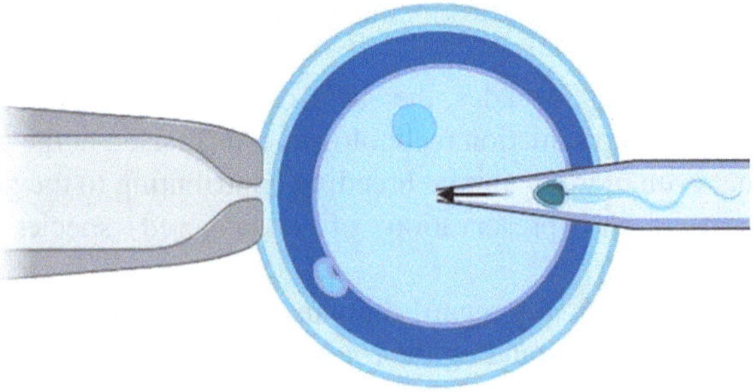

Fig. 1.5 Simplified representation of the Intracytoplasmic Sperm Injection procedure. A spermatozoon is selected and injected into the oocyte held by a holding pipette

natural ratio of many spermatozoa to one egg, establishing a 1:1 relationship between oocyte and spermatozoon. In case of infertile men, a question arises: is it better to try to improve the fertilizing potential of a man or just use a spermatozoon chosen by an operator under a light microscope and try ICSI? Is it still worth studying this amazing cell as so many scientists have done over the centuries?

These techniques are without any doubts one of the major medical advancements; however, as biologists it should be considered that they circumvent reproductive barriers and changes selection pressures, allowing the reproduction of individuals that are not able to reproduce naturally.

For example, IVF favors spermatozoa that swim fast over short distances, whereas coitus benefits spermatozoa with forward motility over longer distances (Hanevik and Hessen 2022); in the ICSI procedure, the natural reproductive barriers are totally overcome because the operator choses and picks-up the single sperm to inject. Since IVF and ICSI are both young technologies, it is difficult to argue the potential longer-term impacts.

In any case, the advent of ART opened the way to the assessment of basic human semen parameters. For this reason, in 1980 the World Health Organization (WHO) published the first "WHO Laboratory Manual for the Examination and Processing of Human Semen" with the declared aim to fulfill the growing needs for the standardization of semen analysis procedures (WHO 1980). Since then the manual has been constantly updated and enriched with validated tests to provide a functional assessment of the spermatozoa, not only the measure of the standard parameters that have been revised in each edition; the last edition (sixth edition) of WHO manual was released in 2021 (Wang et al. 2022).

Even though the evaluation of standard semen parameters represents the first line choice in checking the possible fertilizing potential of an ejaculate, it is not predictable of the chance of natural pregnancy as well as of ART outcome. The approach to semen testing used in the clinical practice is still lacking the best method to optimize the sperm quality in view of ART procedures. For these reasons, numbers of sperm functional tests have been proposed that often are part of basic research tools and are excluded from the clinical routine semen testing (Oehninger et al. 2014).

1.5 The Spermatozoon from Genetics to New Acquisitions Influencing Reproductive Outcome

In the second half of twentieth century, a great interest increased around cytogenetic studies and human spermatozoa. These studies gave great information on the impact and consequences of somatic chromosomal abnormalities on human spermatogenesis (Shi and Martin 2001). In that period, it became clear that infertile individuals with normal karyotype had an increased risk to father a chromosomally abnormal child (Martin 2008). Many research groups started to analyze directly the sperm chromosomes by fluorescence in situ hybridization (FISH) using probes, complementary to DNA sequences, labeled with different fluorochromes (Robbins et al. 1993; Egozcue et al. 1997). It soon became clear that FISH provided a promising method to calculate the frequency of disomic and diploid spermatozoa in fertile men, in infertile patients (Sun et al. 2006; Collodel et al. 2007), in men with altered somatic karyotype (Baccetti et al. 2005; Benet et al. 2005). In the same period, it was observed that the presence of microdeletions in particular regions of Y chromosome were associated with spermatogenic failure (McElreavey et al. 1999) and the PCR screening for Y microdeletion was included in the clinical practice of andrology laboratory (Krausz et al. 2000; Simoni et al. 2004).

Recently, Schlegel in 2023 presented the most important areas of research concerning sperm assessment in the prediction of the reproductive outcome. Essentially, they included the study of sperm DNA fragmentation that is thought to be the most predictive parameter, the assessment of sperm-specific phospholipase C (PLC)ζ, the factor needed in oocyte activation. In addition, the study of sperm centrosome is of particular interest because during fertilization it organizes the mitotic spindle of the oocyte influencing chromosomal segregation, zygotic division, and cytoskeletal organization of the embryo. To this end, new microscopical methods have been recently used to study sperm biology and the impact of sperm morphology on fertilization, such as cryo-electron microscopy, tissue clearing and expansion microscopy, and super-resolution light microscopy (Fishman et al. 2018; Ded et al. 2020; Le Guennec et al. 2020; Wang and Chung 2021). For example, by these technical advancements, the Avidor Reiss group (Fishman et al. 2018) proposed that the distal centriole of human sperm is not reduced during maturation as believed until that moment, but it is remodeled assuming an atypical structure. Despite the variable and atypical structure, the distal centriole can function in the sperm and the zygote (Avidor-Reiss and Fishman 2018). The "centriole remodeling hypothesis," not universally accepted yet, emerged in contrast with the classical "hypothesis of centrosomal reduction."

This discovery related to structure and function of human distal sperm centriole and, in general, the study of the human sperm centrosome is particularly relevant because centrioles are involved in gamete development, fertilization, and embryonic development (Avidor-Reiss and Fishman 2018).

These represent research topics on the edge of major findings that could change our understanding of fertilization and eventually could lead to the development of more efficient diagnoses and strategies in the ART field.

Furthermore, many epigenetic factors involved in gene regulation related to sperm function are potentially important for fertility (McSwiggin and O'Doherty 2018). Much research is devoted to probing the presence and the expression of sperm coding and non-coding RNA (Chan et al. 2022) as well as epigenetic changes, in order to identify potential alterations responsible for impaired spermatogenesis or abnormal sperm function.

In these last two decades, important proteins have been identified in spermatozoa: among them, the sperm-specific ion channel CatSper (Ren et al. 2001) present in the principal piece of sperm tail and involved in sperm motility and Izumo (Inoue et al. 2005), a sperm fusion-related antigen that is reputed to be the sperm membrane protein that recognizes Juno in the oolemma before the fusion between the two gametes.

Another important, quite recent acquisition concerned the discovery that oxidative stress (OS) is one of the major causes of alteration in spermatozoa and is consequently involved in male infertility. The first observation of the correlation between OS and defective spermatozoa was made by MacLeod in 1943; since then the literature has been flooded with comments on this topic. Nowadays, it is accepted that a variety of situations as pathologies, lifestyle, obesity, and exposure to different environmental pollutants can trigger oxidative stress and damage spermatozoa causing lipid peroxidation, DNA fragmentation and negatively influencing the motility (Aitken et al. 2022; Collodel et al. 2022).

1.6 Science Fiction or Reality? From Spermatogenesis In Vitro to Algorithms of Artificial Intelligence and Virtual Reality

A milestone, which was a dream until about a decade ago, was the in vitro production of functional spermatozoa achieved in mice by Sato's group in 2011 (Sato et al. 2011).

Thus, this group of researchers showed that the appropriated and specific conditions of the organ culture could support all steps of the mouse spermatogenesis in vitro in the absence of a circulatory system.

Spermatogenesis is one of the most intricate and lengthy processes of progressive mitotic and meiotic cell division and differentiation, so the production of spermatozoa in vitro has required substantial and prolonged effort to replicate this process.

Therefore, this discovery was acclaimed with great enthusiasm because it represents a potential advancement in the treatment of male infertility, and also an important in vitro model to test the effects of different compounds on spermatogenesis.

During the past decade, scientists struggled in developing a micro-environment able to support spermatogenesis in vitro and several culture strategies have been applied such as the use of various stem cells, ex vivo cultures of testicular tissue (Saitou and Hayashi 2021; Lei et al. 2023), tubule culture, cell culture, and 3-dimensional cell culture methods (Komeya et al. 2018). The recent development of organoid or 3-dimensional culture techniques seems encouraging, but they still need further improvements. It is clear that other studies are needed in order to guarantee the correctness of all meiotic events that are articulated by the checkpoints and to obtain a complete spermiogenesis, during which round spermatids differentiate into elongated spermatids; this last step has not been yet achieved in vitro by various cell culture strategies (Lei et al. 2023). Moreover, the success of mouse spermatogenesis is yet to be applied to other animals (Komeya et al. 2018).

Now we are living a particular period in which we are coming to terms with artificial intelligence (AI), the science based on algorithms able to solve problems or perform tasks and actions typical of the human mind, such as visual perception, speech recognition, decision-making, and translation between languages. AI is now a reality that is revolutionizing many fields, including medicine, and, in this particular context, it can also find interesting applications in the field of medically assisted reproduction. AI algorithms, implemented with deep learning algorithms, can be used in the IVF laboratory to provide fast, unbiased, and accurate evaluations of clinical parameters and procedures including cell tracking, embryo transfer selection, micromanipulation, oocyte, fertilization and embryo assessment, and spermatozoa selection (Jiang and Bormann 2023).

Since it is known that standard evaluation of semen parameters has poor predictive value for reproductive potential and ART outcome, AI algorithms can show the potential to address the challenges associated with sperm selection due to their high objectivity and large-data processing capabilities (Cherouveim et al. 2023). Overall, AI holds much promise for enhancing not only ART clinical outcome, but also laboratory procedures.

In addition, the algorithms utilized for sperm evaluation, applied to hundreds of millions of images, are promising in predicting the reproductive competence of the single spermatozoon: this is a critical process for sperm selection before ICSI (Schlegel 2023).

Finally, the relationship between spermatozoa and virtual reality (VR) is worth mentioning. VR is a computer-generated artificial simulation of a three-dimensional environment that an individual may interact with, using specialized electronic equipment as gloves with sensor or a headset, in a way that he is immersed in an apparent real and tangible situation.

Recently, Rosenkjær et al. (2022) used VR headsets to present erotica during masturbation of a group of donors and found that VR immersion played a strong positive effect on the number of motile sperm in ejaculates of individuals that had an abstinence time shorter than 24 h. They suggested that VR use could improve sperm parameters of infertile men and of cancer patients depositing sperm before treatment to be cryopreserved for future use.

Amazingly, human spermatozoa have become protagonists of a VR game with the idea to learn by playing (Collodel et al. 2019). At the University of Siena, the Department of Political, Social and Cognitive Sciences in partnership with the Department of Molecular and Developmental Medicine developed the VR game "Oxistress" (a sample is visible here: Oxistress—YouTube) aiming at demonstrating the effects of appropriate nutrition on spermatozoa. The game is settled in a spaceship and the aim is the control of sperm quality in a group of astronauts that must colonize a new planet. The player is required to choose different foods or combination of foods and then discover their effects on virtual spermatozoa. Immersed in the epididymis during gameplay, the player may observe how dietary n-3 fatty acids may change the sperm plasma membrane and, as a result, modify the characteristics of the spermatozoa. In summary, this novel approach facilitates the understanding of the critical role that nutrients play in sperm quality and, therefore in male reproduction.

The VR simulation didactic game might serve as a tool to increase public awareness of scientific research on the benefits of healthy eating practices for human reproduction and wellness. In addition, offering a realistic, lifelike environment, the game increases the participation of the subjects implementing the learning process.

1.7 Conclusion

From Leeuwenhoek's *animalcula* to spermatozoa protagonists of the virtual reality, the research has made incredible breakthroughs and it is still extremely flourishing, therefore much remains to be discovered about these fascinating and magnificent cells.

In our journey through the history of spermatozoa, we have met many important figures and seen how many attempts were made to provide explanations to natural phenomena concerning reproduction. This written text suggests us the crucial importance of recollecting history; in this way, we can follow the development of the scientific discoveries and the thinking, and we perceive the possibility to solve old problems with the new technologies now available.

Acknowledgments We are grateful to the Prof. Baccio Baccetti who transmitted to us the curiosity and knowledge of studying the history of the spermatozoon with his enthusiasm in approaching sciences.

References

Adlam JP (1980) Artificial insemination. British. J Sex Med 56:10–14

Afzelius BA, Baccetti B (1991) History of spermatology. In: Baccetti B (ed) Comparative spermatology 20 years after. Raven Press, New York, pp 1–10

Agarwal A, Mulgund A, Hamada A, Chyatte MR (2015) A unique view on male infertility around the globe. Reprod Biol Endocrinol 13:37. https://doi.org/10.1186/s12958-015-0032-1

Aitken RJ, Drevet JR, Moazamian A, Gharagozloo P (2022) Male infertility and oxidative stress: a focus on the underlying mechanisms. Antioxidants (Basel) 11(2):306. https://doi.org/10.3390/antiox11020306

Avidor-Reiss T, Fishman EL (2018) It takes two (centrioles) to tango. Reproduction 157:R33–R51

Baccetti B, Collodel G, Marzella R, Moretti E, Piomboni P, Scapigliati G, Serafini F (2005) Ultrastructural studies of spermatozoa from infertile males with Robertsonian translocations and 18, X. Y aneuploidies Hum Reprod 20(8):2295–2300. https://doi.org/10.1093/humrep/dei050

Benet J, Oliver-Bonet M, Cifuentes P, Templado C, Navarro J (2005) Segregation of chromosomes in sperm of reciprocal translocation carriers: a review. Cytogenet Genome Res 111(3–4):281–290. https://doi.org/10.1159/000086901

Birkhead TR (2000) Promiscuity: an evolutionary history of sperm competition. Harvard University Press, Cambridge, MA, USA

Birkhead TR, Montgomerie R (2009) Three centuries of sperm research. In: Birkhead TR, Hosken DJ, Pitnick SS (eds) Sperm biology: an evolutionary perspective. Elsevier, London, pp 1–42

Castellani C (1973) Spermatozoan biology from Leeuwenhoek to Spallanzani. J Histo Biol 6(1):37–68. https://www.jstor.org/stable/4330587

Chan SY, Wan CWT, Law TYS, Chan DYL, Fok EKL (2022) The sperm small RNA transcriptome: implications beyond reproductive disorder. Int J Mol Sci 23(24):15716. https://doi.org/10.3390/ijms232415716

Chemes HE (2018) Phenotypic varieties of sperm pathology: genetic abnormalities or environmental influences can result in different patterns of abnormal spermatozoa. Anim Reprod Sci 194:41–56. https://doi.org/10.1016/j.anireprosci.2018.04.074

Chemes HE, Alvarez Sedo C (2012) Tales of the tail and sperm headaches: changing concepts on the prognostic significance of sperm pathologies affecting the head, neck and tail. Asian J Androl 14(1):14–23. https://doi.org/10.1038/aja.2011.168

Chemes EH, Rawe YV (2003) Sperm pathology: a step beyond descriptive morphology. Origin, characterization and fertility potential of abnormal sperm phenotypes in infertile men. Hum Reprod Update 9(5):405–428. https://doi.org/10.1093/humupd/dmg034

Cherouveim P, Velmahos C, Bormann CL (2023) Artificial intelligence for sperm selection-a systematic review. Fertil Steril 120(1):24–31. https://doi.org/10.1016/j.fertnstert.2023.05.157

Cobb M (2006) Heredity before genetics: a history. Nat Rev Genet 7:953–958. https://doi.org/10.1038/nrg1948

Collodel G, Capitani S, Baccetti B, Pammolli A, Moretti E (2007) Sperm aneuploidies and low progressive motility. Hum Reprod 22(7):1893–1898. https://doi.org/10.1093/humrep/dem099

Collodel G, Masini M, Signorini C, Moretti E, Castellini C, Noto D, Mattioli S, Santalucia V, Baraldi S, Lusuardi LM, Sirizzotti M, Innocenti A (2019) Antioxidants, dietary fatty acids, and sperm: a virtual reality applied game for scientific dissemination. Oxidative Med Cell Longev 2019:2065639. https://doi.org/10.1155/2019/2065639

Collodel G, Moretti E, Noto D, Corsaro R, Signorini C (2022) Oxidation of polyunsaturated fatty acids as a promising area of research in infertility. Antioxidants (Basel) 11(5):1002. https://doi.org/10.3390/antiox11051002

Ded L, Hwang JY, Miki K, Shi HF, Chung JJ (2020) 3D in situ imaging of the female reproductive tract reveals molecular signatures of fertilizing spermatozoa in mice. elife 9:e62043

Egozcue J, Blanco J, Vidal F (1997) Chromosome studies in human sperm nuclei using fluorescence in-situ hybridization (FISH). Hum Reprod Update 3(5):441–452. https://doi.org/10.1093/humupd/3.5.441

Fishman EL, Jo K, Nguyen Q, Kong D, Royfman R, Cekic AR, Khanal S, Miller AL, Simerly C, Schatten G, Loncarek J, Mennella V, Avidor-Reiss T (2018) A novel atypical sperm centriole is functional during human fertilization. Nat Commun 9:2210

Fitzpatrick JL, Kahrl AF, Snook RR (2022) SpermTree, a species-level database of sperm morphology spanning the animal tree of life. Sci Data 9(1):30. https://doi.org/10.1038/s41597-022-01131-w

Hanevik HI, Hessen DO (2022) IVF and human evolution. Hum Reprod Update 28(4):457–479. https://doi.org/10.1093/humupd/dmac014

Howards SS (1997) Antoine van Leeuwenhoek and the discovery of sperm. Fertil Steril 67(1):16–17. https://doi.org/10.1016/s0015-0282(97)81848-1

Inoue N, Ikawa M, Isotani A, Okabe M (2005) The immunoglobulin superfamily protein Izumo is required for sperm to fuse with eggs. Nature 434(7030):234–238. https://doi.org/10.1038/nature03362

Jiang VS, Bormann CL (2023) Artificial intelligence in the in vitro fertilization laboratory: a review of advancements over the last decade. Fertil Steril 120(1):17–23. https://doi.org/10.1016/j.fertnstert.2023.05.149

Kimura Y, Yanagimachi R (1995) Intracytoplasmic sperm injection in the mouse. Biol Reprod 52:709–720

Komeya M, Sato T, Ogawa T (2018) In vitro spermatogenesis: a century-long research journey, still halfway around. Reprod Med Biol 17(4):407–420. https://doi.org/10.1002/rmb2.12225

Krausz C, Quintana-Murci L, McElreavey K (2000) Prognostic value of Y deletion analysis: what is the clinical prognostic value of Y chromosome microdeletion analysis? Hum Reprod 15(7):1431–1434. https://doi.org/10.1093/humrep/15.7.1431

Le Guennec M, Klena N, Gambarotto D, Laporte MH, Tassin AM, van den Hoek H, Erdmann PS, Schaffer M, Kovacik L, Borgers S, Goldie KN, Stahlberg H, Bornens M, Azimzadeh J, Engel BD, Hamel V, Guichard P (2020) A helical inner scaffold provides a structural basis for centriole cohesion. Sci Adv 6:eaaz4137

Lei Q, van Pelt AMM, Hamer G (2023) In vitro spermatogenesis: why meiotic checkpoints matter. Curr Top Dev Biol 151:345–369. https://doi.org/10.1016/bs.ctdb.2022.04.009

Levine H, Jørgensen N, Martino-Andrade A, Mendiola J, Weksler-Derri D, Jolles M, Pinotti R, Swan SH (2023) Temporal trends in sperm count: a systematic review and meta-regression analysis of samples collected globally in the 20th and 21st centuries. Hum Reprod Update 29(2):157–176. https://doi.org/10.1093/humupd/dmac035

Martin RH (2008) Cytogenetic determinants of male fertility. Hum Reprod Update 14(4):379–390. https://doi.org/10.1093/humupd/dmn017

McElreavey K, Krausz C, Bishop CE (1999) The human Y chromosome and male infertility. In: McElreavey K (ed) The genetic basis of male infertility. Springer, Heidelberg, Germany, p 211

McSwiggin HM, O'Doherty AM (2018) Epigenetic reprogramming during spermatogenesis and male factor infertility. Reproduction 156(2):R9–R21. https://doi.org/10.1530/REP-18-0009

Mortimer ST, van der Horst G, Mortimer D (2015) The future of computer-aided sperm analysis. Asian J Androl 17(4):545–553. https://doi.org/10.4103/1008-682X.154312

Oehninger S, Franken DR, Ombelet W (2014) Sperm functional tests. Fertil Steril 102(6):1528–1533. https://doi.org/10.1016/j.fertnstert.2014.09.044

Palermo G, Joris H, Devroey P, Van Steirteghem AC (1992) Pregnancies after intracytoplasmic injection of single spermatozoon into an oocyte. Lancet 340:17–18

Palermo GD, Neri QV, Monahan D, Kocent J, Rosenwaks Z (2012) Development and current applications of assisted fertilization. Fertil Steril 97:248–259. https://doi.org/10.1016/j.fertnstert.2011.12.037

Puerta Suárez J, du Plessis SS, Cardona Maya WD (2018) Spermatozoa: a historical perspective. Int J Fertil Steril 12(3):182–190. https://doi.org/10.22074/ijfs.2018.5316

Pypłacz J (2022) All about my mother… Cilissa, apollo and the Unity of the Oresteia. Studia Litteraria Universitatis Iagellonicae Cracoviensis 17:241–253

Ren D, Navarro B, Perez G, Jackson AC, Hsu S, Shi Q, Tilly JL, Clapham DE (2001) A sperm ion channel required for sperm motility and male fertility. Nature 413(6856):603–609. https://doi.org/10.1038/35098027

Robbins WA, Segraves R, Pinkel D, Wyrobek AJ (1993) Detection of aneuploid human sperm by fluorescence in situ hybridization: evidence for a donor difference in frequency of sperm disomic for chromosomes 1 and Y. Am J Hum Genet 52(4):799–807

Rosenkjær D, Pacey A, Montgomerie R, Skytte AB (2022) Effects of virtual reality erotica on ejaculate quality of sperm donors: a balanced and randomized controlled cross-over within-subjects trial. Reprod Biol Endocrinol 20(1):149. https://doi.org/10.1186/s12958-022-01021-1

Rowe M, Rosengrave P (2020) There and back again: a sperm's tale. Mol Reprod Dev 87(4):395–398. https://doi.org/10.1002/mrd.23325

Saitou M, Hayashi K (2021) Mammalian in vitro gametogenesis. Science 374(6563):eaaz6830. https://doi.org/10.1126/science.aaz6830

Sandler I (1973) The re-examination of Spallanzani's interpretation of the role of the spermatic animalcules in fertilization. J Hist Biol 6(2):193–223. http://www.jstor.org/stable/4330595

Sato T, Katagiri K, Gohbara A, Inoue K, Ogonuki N, Ogura A, Kubota Y, Ogawa T (2011) In vitro production of functional sperm in cultured neonatal mouse testes. Nature 471(7339):504–507

Schlegel PN (2023) Can sperm quality affect reproductive outcomes? Fertil Steril 120(4):707–708. https://doi.org/10.1016/j.fertnstert.2023.08.007

Shi Q, Martin RH (2001) Aneuploidy in human spermatozoa: FISH analysis in men with constitutional chromosomal abnormalities, and in infertile men. Reproduction 121(5):655–666. https://doi.org/10.1530/rep.0.1210655

Simoni M, Bakker E, Krausz C (2004) EAA/EMQN best practice guidelines for molecular diagnosis of Y-chromosomal microdeletions. State of the art 2004. Int J Androl 27:240–249

Steptoe PC, Edwards RG (1978) Birth after reimplantation of a human embryo. Lancet 312:366

Sun F, Ko E, Martin RH (2006) Is there a relationship between sperm chromosome abnormalities and sperm morphology? Reprod Biol Endocrinol 4:1. https://doi.org/10.1186/1477-7827-4-1

Uehara T, Yanagimachi R (1976) Microsurgical injection of spermatozoa into hamster eggs with subsequent transformation of sperm nuclei into male pronuclei. Biol Reprod 15:467–470

Wang H, Chung JJ (2021) Mitochondria-cytoskeleton interactions in mammalian sperm revealed by cryoelectron tomography. Proc Natl Acad Sci USA 118(48):e2118020118. https://doi.org/10.1073/pnas.2118020118

Wang C, Mbizvo M, Festin MP, Björndahl L, Toskin I, other Editorial Board Members of the WHO Laboratory Manual for the Examination and Processing of Human Semen (2022) Evolution of the WHO "semen" processing manual from the first (1980) to the sixth edition (2021). Fertil Steril 117(2):237–245. https://doi.org/10.1016/j.fertnstert.2021.11.037

World Health Organization (1980) Laboratory manual for the examination of human semen and semen-cervical mucus interaction. Press Concern, Singapore

Yanagimachi R (1981) Mechanisms of fertilization in mammals. In: Mastroianni L, Biggers JD (eds) Fertilization and embryonic development in vitro. Boston, MA, USA, Springer Science and Business Media

Human Spermatogenesis

2

Contents

Core Messages

This chapter is mainly focused on the presentation of spermatogenesis in humans. First, we addressed the history of the acquisition of knowledge of spermatogenesis, the characterization of Sertoli and Leydig cells, and of the blood–testis barrier. In the second part, the different phases of the spermatogenesis, the mitotic and the meiotic processes, are presented with references to recent acquisitions of the role of several proteins. Finally, the process of spermiogenesis including the chromatin condensation, the formation of the acrosome, and the flagellum with the axoneme, its accessory structures and the mitochondria are described.

Abbreviations

ABP	Androgen binding protein
Bmp4	Bone morphogenetic protein 4
BTB	Blood–testis barrier
FSH	Follicle stimulating hormone
GDNF	Glial cell line-derived neurotrophic factor
INSL3	Insulin-like peptide 3

E. Moretti, G. Collodel, *The Human Spermatozoon*,
https://doi.org/10.1007/978-3-031-97868-5_2

LH	Luteinizing hormone
LIF	Leukemia inhibitory factor
PModS	Peritubular factor that modulates Sertoli cell function
SRY	Sex-determining region Y
StAR	Steroidogenic acute regulatory protein
TGF-β	Transforming growth factor

2.1 A Brief Overview of the Spermatogenesis Discovery

It is hard to discuss spermatozoa without mentioning the environment in which they develop.

Spermatogenesis is a complex biological process that produces haploid spermatozoa from diploid spermatogonial stem cells within seminiferous tubule boundaries of the testis.

The first person who described and illustrated the thin tubules located in the human testis was Claude Aubery (1658), although their important role in spermatogenetic process was still unknown. As reported in Chap. 1 (see Sect. 1.1), Leeuwenhoek first observed spermatozoa (1677) only hypothesizing that they originated from testicular tissue; much later, Kolliker (1841) discovered in marine invertebrate species the testicular origin of spermatozoa, creating a *trait d'union* between Leeuwenhoek and Aubery (Birkhead and Montgomerie 2009).

In the mid-nineteenth century, Sir Astley Cooper, a British surgeon, and anatomist, published a book entitled "Observations on the structure and diseases of the testis" that reported the site of spermatozoa origin (Cooper 1845).

Speaking of the history of the discovery of spermatogenesis and of their main protagonists, we must mention Enrico Sertoli (1842–1910), an Italian histologist and physician devoted to the study of a wide variety of topics including spermatogenesis. He described, in the seminiferous epithelium, a type of characteristic cell that supports and nurtures with their extensive cytoplasm the developing germinal cells through the spermatogenetic process: Sertoli cells (Sertoli 1865).

At that time, a major debate among scientists of reproductive field concerned the origin of spermatozoa. Victor von Ebner supported the view that spermatozoa originated from Sertoli cells, whereas the germ cells degenerated producing seminal plasma. In contrast, Enrico Sertoli's group held the opinion (we now know that it is correct) that male germ cells progress through a sequential developmental program that results in the production of spermatozoa, suggesting a supporting role of Sertoli cells (Geyer 2018).

In 1850, Franz von Leydig, a German zoologist and comparative anatomist, first described the eponymous testicular cells (Leydig 1850) located in the interstitial tissue that are now known to be the main source of testosterone, essential for spermatogenesis and sexual differentiation.

From that moment, much of the work on spermatogenesis concerned the detailed description of the process, and the estimate of its duration in different organisms. Soon after, several studies on hormonal regulation of spermatogenesis emerged in the scientific scenario and androgens were identified as the most important hormones able to orchestrate spermatogenesis (Walsh et al. 1934).

2.2 Introducing the Sperm Production Process

In the kingdom Animalia, gametogenesis, the production of germinal cells, occurs in a rather varied manner depending on the species. This chapter is mainly focused on the presentation of spermatogenesis in humans.

The seminiferous tubules represent the operational unit of the testis and are composed by three types of cells: the peritubular myoid cells, Sertoli cells, and the germ cells at different stages of maturation (Fig. 2.1). Adjacent to seminiferous tubules, in the interstitial tissue, clusters of Leydig cells represent the primary source of testosterone.

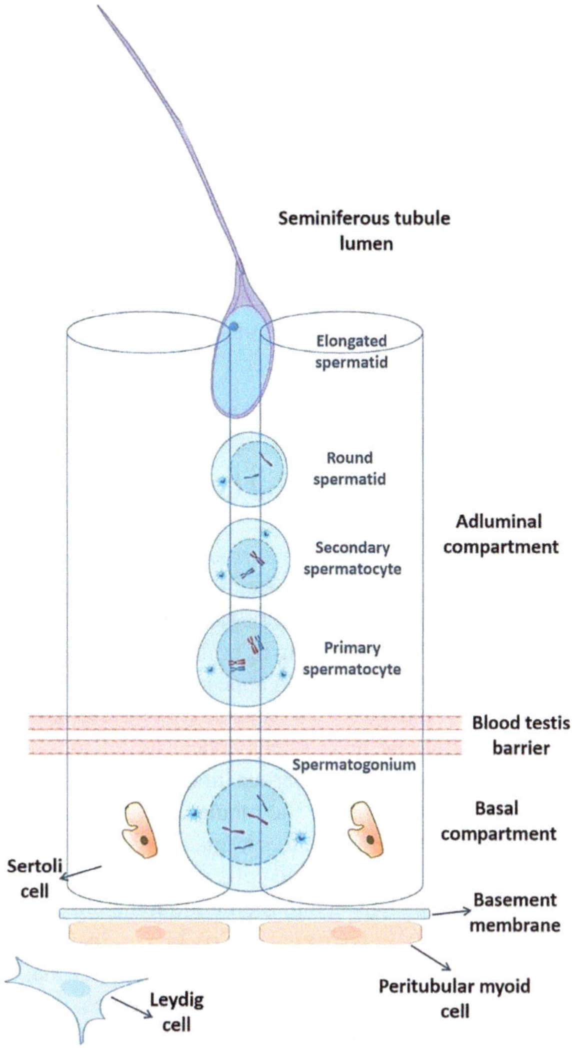

Fig. 2.1 Simplified representation of a section of seminiferous tubule surrounded by peritubular myoid cells. Inside the tubule, Sertoli cells with their wide cytoplasm envelop all the cell stages of the spermatogenesis and spermiogenesis (spermatogonia, primary spermatocytes, secondary spermatocytes, round spermatids, and elongated spermatids). The blood–testis barrier is a complex structure arranged by Sertoli cells that form two compartments: the basal one, with spermatogonia and preleptotene spermatocytes and the adluminal one with spermatocytes and post-meiotic germ cells

Spermatogenesis is the process by which, after puberty, the unipotent stem cells, the spermatogonia, localized near to the basement membrane of the testicular seminiferous epithelium, undergo different steps of proliferative phases and maturation processes that culminate in the production of male gametes, the spermatozoa (Fig. 2.1). The whole process includes the proliferation of spermatogonia, the spermatogonia differentiation into spermatocytes, the meiotic divisions of spermatocytes that produce the spermatids, the maturation of early spermatids, round cells rich in cytoplasm, into spermatozoa that are released in the testis lumen.

At the end of this process, the spermatozoa undergo a further maturation in epididymis and mix with seminal plasma produced by the prostate gland, seminal vesicles, and bulbourethral glands; finally, they are expelled from the male body.

In 1963, Heller and Clermont suggested that, in humans, one cycle of the seminiferous epithelium lasts 16 days, and the whole spermatogenesis is estimated to consume approximately 74 days. However, a more recent study on normozoospermic men reported that the spermatogenetic cycle takes a mean of 64 ± 8 days (range 42–76): in one case, the time lag was 42 days but in all other cases the spermatogenesis required at least 60 days (Misell et al. 2006).

2.3 Somatic Cells Involved in Spermatogenesis

Spermatogenesis is a process that requires a specific regulation and cross-talking between somatic cell populations, including peritubular myoid cells, Leydig cells, and Sertoli cells. Below, a general and essential view of the function of these somatic cells is presented.

2.3.1 Peritubular Myoid Cells

The peritubular myoid cells surround the seminiferous tubules in the testis and are present in all mammalian species, suggesting a role as an integral and functional component of the testis (Maekawa et al. 1996; Verhoeven et al. 2000).

In humans, peritubular myoid cells are arranged in several irregular layers around the seminiferous tubules and appear as thin, flat, spindle-shaped cells with smooth-muscle-like characteristics embedded in the extracellular matrix (Fig. 2.2).

The peritubular myoid cells appear to collaborate with Sertoli cells in building the basal lamina of seminiferous tubules, providing a specific niche for spermatogonia stem cells that inside the tubule can mitotically divide to maintain the pool of stem cells or differentiate to spermatocytes and spermatids. The most intuitive and important function of myoid cells regards the propulsion of immotile spermatozoa toward the rete testis due to their contractile ability, using the actin-myosin system (Mayerhofer 2013; Neto et al. 2016). One of the studied regulators of contraction of human peritubular myoid cells is angiotensin II that acts via AT1 receptor (Welter et al. 2014). Beside these contractile properties, the peritubular myoid cells participate in the organization of testicular functions due to their strategic location at the

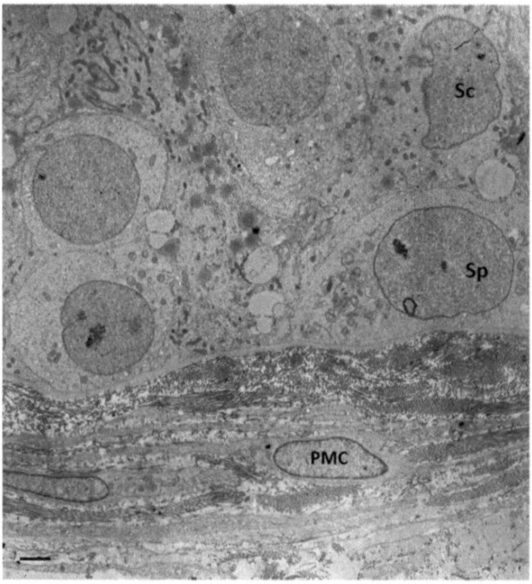

Fig. 2.2 Transmission electron microscopy micrograph of a section of human testis seminiferous tubule. The tubule is surrounded by flat, thin cells: the peritubular myoid cells (PMC). Pale A spermatogonia are located on the basal lamina of the seminiferous tubule (Sp). A Sertoli cell (Sc), with the typical irregularly shaped nucleus, is visible in the upper right part of the picture. Bar: 1 µm

edge between interstitial compartment where Leydig cells are housed, and tubular compartment where germ cells and Sertoli cells are located. For this reason, many studies have demonstrated the paracrine influences exerted by peritubular cells on somatic cells secreting a number of interleukins, growth factors, extracellular matrix components as fibronectin, type I and IV collagens, proteoglycans (for a complete view of the issue the following reviews can be consulted: Mayerhofer 2013; Zhou et al. 2019).

One of the most intriguing paracrine factors secreted by peritubular myoid cells is the PModS, which means "Peritubular factor that Modulates Sertoli cell function" (Verhoeven et al. 2000). Some evidence suggests that PModS plays an indirect role in spermatogenesis modulating Sertoli cells differentiation and function acting on both inhibin secretion, androgen responsiveness and secretion of transferrin (Neto et al. 2016). Another important factor secreted by peritubular myoid cells is the leukemia inhibitory factor (LIF) that influences the formation of blood–testis barrier (BTB) made by Sertoli cells interactions, assuring the integrity of seminiferous tubule structure.

These cells also secrete Glial cell line-derived neurotrophic factor (GDNF) that is a crucial growth factor involved in the self-renewal of spermatogonial stem cells (Chen et al. 2016).

On the other hand, different compounds regulate the activity of peritubular myoid cells that are an androgen target since they express androgen receptor as Leydig and Sertoli cells, but in a lesser extent. A mouse animal model gene deleted for androgen receptor in peritubular myoid cells (PTM-specific AR-knockout mice) indicates that androgens play a pivotal role in the modulation of peritubular myoid cells physiological function (Welsh et al. 2009).

Considering all these studies, it is evident that peritubular myoid cells provide structural integrity to the seminiferous tubule; in addition, they participate in the regulation of spermatogenesis and the testicular function. In the field of spermatogenesis studies, these cells probably received less attention and deserve many efforts to understand the real role played as modulator of the spermatogenetic process.

2.3.2 Leydig Cells

Leydig cells are located, as either single cells or clusters, within the interstitial space (connective tissue), in a strategic position between the seminiferous tubules and the blood vessels. Focusing on the development of Leydig cells, three stages can be observed: the proliferation of Leydig cell precursor, the differentiation in immature and finally adult Leydig cells (Mendis-Handagama and Ariyaratne 2001). This triphasic development is concomitant with the three waves of testosterone: at 14–18 weeks before birth, 2–3 months after birth, and from puberty throughout adult life (Prince 2001; Zirkin and Papadopoulos 2018).

Human Leydig cells are irregular in shape and show characteristics compatible with their steroidogenic activity: wide smooth endoplasmic reticulum, numerous mitochondria, and lipid droplets. The hallmark of these cells is the presence of the enigmatic Reinke crystals, which are thought to be of protein origin, even though the molecular composition has not yet been fully defined (Planinić et al. 2022).

The most famous activity of Leydig cells regards the production of androgens that diffuse into the seminiferous tubules and blood vessels present in the interstitium. Androgens lead the signaling pathways controlling spermatogenesis via androgen receptors that are present mainly on Sertoli cells and peritubular myoid cells (Zhou et al. 2019), in addition, testosterone and dihydrotestosterone play a pivotal role in male sexual differentiation and deeply influence sexual and psychological behavior of males (Corona et al. 2022).

The production and release of testosterone depends on stimulation of Leydig cells by the luteinizing hormone (LH), produced by the pituitary gland, via LH receptor on the Leydig cell membrane. This pathway triggers a complex signaling that activates the protein "steroidogenic acute regulatory protein" StAR that transports cholesterol, mobilized from lipid droplets present in the cytoplasm and from plasma membrane, into the mitochondrial inner membrane where it is converted in pregnenolone. The pregnenolone reaches the smooth endoplasmic reticulum and, by multiple reactions, is transformed in testosterone (Zhou et al. 2019). By this mechanism, Leydig cells produce 3–10 mg/mL of testosterone daily, an amount that represents the vast majority of circulating testosterone in adult men. It was believed that the high concentration of intra-testicular testosterone, 100-fold greater than that found in normal human serum (Jarow et al. 2001), was important to sustain the spermatogenetic process. However, Oduwole et al. (2021) have recently revised this dogma demonstrating that high intra-testicular testosterone levels greatly exceed the need for androgen receptor activation and are not required for spermatogenesis. However, they are necessary to produce sufficient circulating testosterone for the development of male secondary sex characteristics as well as metabolic and central actions of sex steroids.

Leydig cells are also the predominant testicular source of estrogens after puberty and the balance between estrogens and androgens is essential for the normal function of male reproduction. The aromatase, the key enzyme responsible for conversion of androgens to estrogens, is expressed mainly in Leydig cells but can be also found in Sertoli cells, germ cells and spermatozoa (Lambard et al. 2003, 2005). This is a key indication on the importance of estrogens in orchestrating the spermatogenetic process.

A number of other studies have found that Leydig cells regulate spermatogenesis, influencing the secretion of paracrine factors such as oxytocin, interleukins, adipocytokines, inhibin, insulin-like growth factors 1, insulin-like peptide 3 (INSL3), and others (Zhou et al. 2019).

In particular, INSL3 is a circulating hormone that is thought to be a biomarker of Leydig cell function (Facondo et al. 2020) more reliable than testosterone itself. Beside its action in mediating the migration of the testis into the scrotum, INSL3 can significantly inhibit the apoptosis of male germ cells, and serum INSL3 content has been shown to be positively correlated with sperm production (Bay et al. 2005).

At present, our knowledge of factors involved in the intracellular signaling cascades that control the development, and function of the human Leydig cells is still incomplete and further studies are needed to fill the gaps of knowledge in this intricate process.

2.3.3 Sertoli Cells

Sertoli cells are the unique non-dividing somatic cells present in the microenvironment of the seminiferous epithelium (Figs. 2.1 and 2.2), active during reproductive lifetime.

Their pivotal function starts during testis development throughout fetal period. From the sixth week of gestation, the gene sex-determining region Y (SRY, Sinclair et al. 1990), located in the short arm of Y chromosome, orchestrates the development of the testis. SRY gene encodes for a transcription factor that upregulates an intricate network of other transcription factors committing the differentiation of Sertoli cells. These cells play an irreplaceable role in the spermatogenetic process, demonstrated by the following facts and observations:

The number of Sertoli cells is correlated with the number of germ cells present in the testes. They occupy 17%–20% of the seminiferous epithelium in adults, and each Sertoli cell can support 30–50 germ cells at different developmental stages (Cheng et al. 2010).

Seminiferous tubules with germ cells only have never been observed, whereas the Sertoli cell-only syndrome, also known as Del Castillo syndrome or germ cell aplasia, is one of the most common causes of azoospermia in infertile men (Leslie et al. 2023).

In the in vitro culture of germ cells, the presence of Sertoli cells is absolutely needed (Baazm et al. 2017).

Sertoli cells are extremely polarized and dynamic cells that change structure during the cycle of the seminiferous epithelium. The base of the cell is in physical contact with the basement membrane and the opposite region points to the lumen of seminiferous tubule (Franca et al. 2016; Griswold 2018). For this reason, one of the main characteristics is represented by the wide cytoplasm that embeds the developing germ cells assisting their maturation and migration toward the epithelium up to the lumen (Neto et al. 2016). These cells contribute to all stages of spermatogenesis from the spermatogonia stem cells to the differentiated spermatozoa.

The study of the hormonal pathway involved in the physiology of Sertoli cells is out of the focus of this manuscript and can be found in Shah et al. (2021). In short, Sertoli cells are the main target for follicle stimulating hormone (FSH). In fact, they express FSH receptor and for testosterone produced by Leydig cells; both FSH and testosterone play a synergic effect on spermatogenesis regulation.

It should be noted that Sertoli cells are believed to be involved also in regulation of other testis somatic cells, including peritubular myoid cell development in the pre-pubertal testis, in supporting the function of the testicular vasculature, and in contributing to testicular immune privilege (O'Donnell et al. 2022). Sertoli cells produce several factors and hormones that participate in orchestrating the cross-talk among somatic and germ cells involved in the spermatogenetic process (Neto et al. 2016; Griswold 2018; O'Donnell et al. 2022).

One of the most cited secretory products is represented by the androgen binding protein (ABP), which is present both in the efferent duct fluid and epididymal fluid (French and Ritzen 1973). ABP is a carrier protein that binds testosterone and dihydrotestosterone with very high affinity, modulating the bioavailability of these androgens in the testis and epididymal environment.

Other proteins secreted by Sertoli cells that contribute to testis microenvironment are: transferrin, ceruloplasmin, vitamin and ion transporters, lactate, acetate, extracellular matrix components, GDNF, transforming growth factor (TGF)-α, TGF-β, interleukins, and others.

It is worth emphasizing that Sertoli cells can drive the germ cells to different pathways depending on the secretion of specific molecules. For example, the pathway involving the production of GDNF leads to spermatogonia renewal and proliferation (Parekh et al. 2019), whereas the differentiation of spermatogonia is concomitant with an increased expression of retinoic acid, stem cell factor, and bone morphogenetic protein 4 (Bmp4) (Thumfart and Mansuy 2023).

In addition, anti-Müllerian hormone, a member of the TGF-β superfamily secreted by immature Sertoli cells from the eighth week of fetal gestation, triggers the regression of the Müllerian ducts and precludes the development of female internal genitalia. Then, Sertoli cells continue to produce anti-Müllerian hormone until puberty and at low level during adulthood, but its function during this period of age remains unclear (Neto et al. 2016; Edelsztein et al. 2022; Lucas-Herald and Mitchell 2022).

Also secreted by Sertoli cells, the inhibin B is a glycoprotein that triggers a negative feedback loop modulating the FSH production from the anterior pituitary gland and positively correlates with Sertoli cell function, sperm number, and spermatogenic status (O'Connor and De Kretser 2004).

While a main source of estrogen in the adult testes is the Leydig cell, Sertoli cells also produce estrogen, in particular in the immature testes (Carreau et al. 2012). Even germ cells seem to contribute significantly to the estradiol synthesis (O'Shaughnessy 2014). Regarding specifically Sertoli cells, the mRNAs for aromatase (enzyme that converts androgen in estrogen) was observed in the cytoplasm at every stage of development.

Due to their high immunogenicity and specific metabolic demands, germ cells need to be kept in a tightly regulated microenvironment generated by the BTB formed by the cell junction between adjacent Sertoli cells. BTB (Fig. 2.1), whose anatomical and functional integrity is essential for spermatogenesis, is located near the base of the seminiferous tubule and defines a physiological compartmentalization of the epithelium into two different areas: the basal one and the adluminal one. The basal compartment hosts spermatogonia and preleptotene spermatocytes, whereas the germ cells undergoing meiosis and spermatids reside in the adluminal compartment. Therefore, the BTB creates a metabolic and immune-privileged microenvironment for cells located in the adluminal compartment and restricts the transit of substances between the two compartments. The aploid cells can express autoantigens recognized by the immune system and need to be protected (Mruk and Cheng 2015). The most important structures of BTB are the cell junctions that connect the adjacent Sertoli cells, as tight junctions, basal ectoplasmic specialization, gap junction, and desmosomes. In particular, tight junctions prevent solutes from crossing the extracellular space and create a specialized environment that protects germ cells from the attack by the host immune system. Ectoplasmic specializations are specific actin-related cell–cell junctions that are present at the BTB level (the basal ectoplasmic specialization), but also between spermatids and Sertoli cells in the adluminal compartment (the apical ectoplasmic specialization) (Mruk and Cheng 2015; Neto et al. 2016).

The first cells to cross the BTB are the preleptotene-primary spermatocytes, which migrate to the adluminal compartment to complete the meiosis. It is intuitive that the junctions present in BTB undergo constant remodeling, by assembly and disassembly, to allow the transport of the preleptotene spermatocytes across the barrier. This is facilitated by internalization of tight junctions -related proteins, so that the BTB is transiently dissolved. In animals, the regulatory mechanisms of the remodeling of junctional proteins that allow the spermatocytes to cross BTB without causing a "leak," remain poorly understood. FSH, androgens and local testicular factors, TGFβ superfamily and retinoid signaling seem to mediate junction assembly and disassembly during the translocation process (Stanton 2016). However, the few human studies covering this topic are not enough to draw conclusions.

Sertoli cells and the developing germ cells they assist are also joined by junctions as ectoplasmic specialization and gap junctions, and carriers that allow the flow of compounds as carbohydrates, lactate, amino acids, vitamins, lipid and ions needed by germ cells.

Finally, the multitasking Sertoli cells play a role as phagocytes as well, cleaning degenerating germ cells and the residual cytoplasm of spermatids. This represents a vital function because a significant fraction of germ cells is eliminated during spermatogenesis. If these dead cells remain in the seminiferous tubules, harmful substances may be released, impairing the sperm production (Nakagawa et al. 2005).

2.4 Overview of Spermatogenesis

Germ cells, the only human cells undergoing the meiotic process, are located with Sertoli cells in the seminiferous epithelium and represent a group of cells meant to become spermatozoa. They are organized in a way by which cells progressively mature from the basal compartment to the luminal one (Fig. 2.1).

The spermatogenetic process includes three functional steps:

(a) Mitotic phase: spermatogonia undergo mitotic division to renew the pool of stem cells, whereas a lower fraction of these cells is committed to start the meiosis, producing primary spermatocytes. These cells are in the basal compartment of the seminiferous epithelium.
(b) Meiotic phase and the production of haploid cells: primary spermatocytes yield secondary spermatocytes by the first meiotic division, and secondary spermatocytes produce spermatids after the second meiotic division. These cells are in the adluminal region of seminiferous epithelium.
(c) Spermiogenesis: the final phase of cell differentiation with the morphological transformation of round spermatids into elongated and flagellated cells: the spermatozoa. This phase takes place in the adluminal region of seminiferous epithelium (Fig. 2.3).

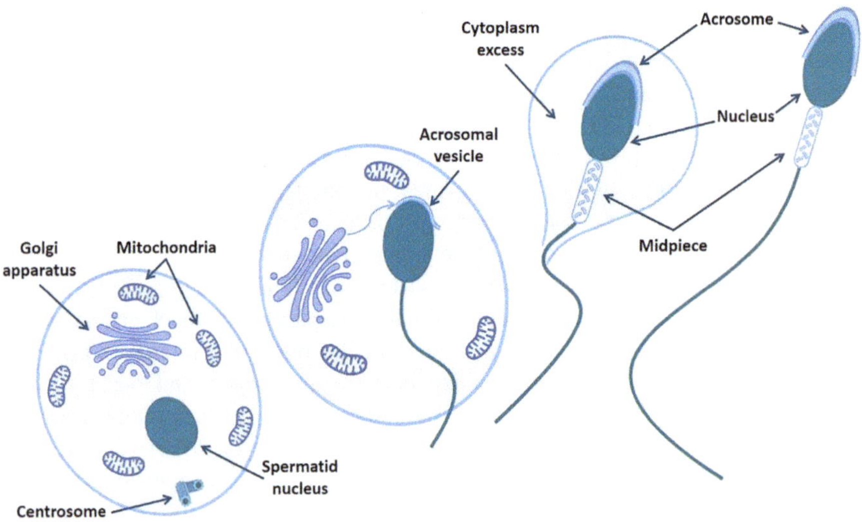

Fig. 2.3 Simplified representation of spermiogenesis: the changes involving the transformation of a round spermatid to a mature spermatozoon are shown. At the end of the meiotic process, the round spermatid is rich in cytoplasm, with a round nucleus and the Golgi complex forming the acrosomal vesicle around the nucleus. In opposite side, the centrosome origins the flagellum. During the development, the acrosome spreads around the nucleus acquiring the typical shape of the species, the chromatin progressively condenses, the mitochondria surround the midpiece, the flagellum elongates, and the cytoplasm is lost

This complex process is under the control of multiple hormones and signaling pathways, among them the hypothalamic–pituitary–testicular axis involving FSH and LH, testosterone and estradiol-17ß are the most important one (O'Donnell et al. 2006; O'Shaughnessy 2014; Chen et al. 2017).

2.4.1 Mitotic Phase of Spermatogenesis

Spermatogonia are diploid progenitors of the other germ cells and are located in the basal compartment of the seminiferous epithelium (Figs. 2.1 and 2.2) near the basement membrane and in close contact with Sertoli cells. Spermatogonia are categorized as A dark spermatogonia, A pale spermatogonia, and B spermatogonia (Clermont 1966), based on the characteristics of the chromatin.

A dark spermatogonia are quite small, roundish cells with uniformly dense chromatin. A pale spermatogonia are larger than A dark spermatogonia, with elongated nuclei with a chromatin showing a granular texture. Nucleoli are visible in both series of spermatogonia.

Many models describing the pattern of spermatogonia proliferation are available (Amann 2008), but in the most accepted and easy one, the stem cells are represented by A dark and A pale spermatogonia.

Due to the low mitotic index, a dark spermatogonia represent a pool of reserve spermatogonia which remain quiescent but can give rise to A pale spermatogonia when needed and are considered to be the "true" spermatogonial stem cells in humans (Chen et al. 2017).

A pale spermatogonia are the active stem germ cells, and they undergo mitosis. The fate of an A pale spermatogonium includes essentially three possibilities: the proliferation for self-renewal, the differentiation in B spermatogonia, committed to proceed with meiotic divisions, and apoptosis. The decision whether the spermatogonia will self-renew, differentiate, or become apoptotic is mainly influenced by Sertoli cells-derived factors.

For example, GDNF promote spermatogonial proliferation (Parekh et al. 2019), whereas cKIT, a transmembrane protein tyrosine kinase receptor, and stem cell factor, its cKIT ligand (Sandlow et al. 1996), are involved, together with Bmp4 and retinoic acid, in the differentiation of human spermatogonia.

In humans, the rate of formation of B spermatogonia is low; therefore, this condition represents a limitation for human spermatogenesis efficiency (Fayomi and Orwig 2018). Morphologically, B spermatogonia, located in the basement membrane, show large sized, clear, roundish nuclei.

However, B spermatogonia, committed for differentiation, loose the contact with the basement membrane and undergo one mitotic division before starting meiosis.

A peculiarity of male spermatogenesis regards the incomplete cell division, the cytokinesis begins but it is arrested, and the daughter cells remain connected by intercellular bridges. These structures provide physical support and synchronous development for the germ cell clones by facilitating the exchange of regulatory compounds (Hermo et al. 2010).

The human spermatogonial compartment is essential for daily production of millions of sperm. Despite this crucial role, the molecular signature, kinetic behavior, and regulation of human spermatogonia are still open questions in reproductive biology (Boitani et al. 2016; Fayomi and Orwig 2018).

2.4.2 Meiotic Phase of Spermatogenesis

The meiosis is a process that in humans starts at puberty. The primary spermatocytes are produced by mitosis of B spermatogonia, they are diploid cells and represent the last germ cells to undergo DNA replication before meiosis. The preleptotene-primary spermatocytes cross the blood–testis barrier and, once reached the adluminal compartment of seminiferous epithelium, start the meiotic process. During prophase of the first meiotic division (meiosis I) different sub-phases can be identified: the chromosomes condense and become evident (leptotene), the homologous chromosomes pair (zygotene) undergo crossing over (pachytene) and finally chiasmata terminalization occurs (diplotene, diakinesis). The crossing over based on DNA recombination is a crucial event both for genetic variability (Grelon 2016) and for a correct chromosome segregation.

Numerous genes and proteins participate in regulating the meiotic DNA recombination, and mutations in these genes trigger germ cell arrest, leading to infertility (Cannarella et al. 2020; Xie et al. 2022). For example, the testis-expressed gene 11 (*TEX11*) is a X-linked meiosis-specific gene encoding a protein involved in the regulation of early-stage germ cells development. TEX11 is an important component of the synaptonemal complex, and its mutations can be associated with non-obstructive azoospermia and maturation arrest (Song et al. 2023). A mouse model mutated for TEX11 shows spermatogenetic impairment due to delayed repair of double strand breaks and reduced crossing over formation in spermatocytes (Adelman and Petrini 2008).

Once crossing over is completed, homologous chromosome segregation occurs and each primary spermatocyte forms two secondary haploid spermatocytes (with haploid number of chromosomes each of which is formed by two sister chromatids), connected by intracellular cytoplasmic bridges that represent important structures for germ cells communication, synchronization, and chromosome dosage compensation in haploid cells.

During meiosis I, a correct chromosome segregation depends upon the formation of connections between homologous chromosome pairs. Crossing overs represent the basis of these connections and chiasmata are the cytological manifestation of crossing overs. At anaphase of meiosis 1, the partners of homologous chromosomes migrate away from one another to opposite poles of the spindle. A protein called shugoshin (Watanabe 2005) is important to protect the centromeric cohesin in meiosis I and to correct wrong kinetochore-microtubule connections (Sun et al. 2022). Shugoshin recruits a protein phosphatase (PP2A) to centromeres, allowing the centromeric cohesin refractory to cleavage by the protease separase at the metaphase I to anaphase I transition (Previato de Almeida et al. 2019).

During anaphase of meiosis II, in secondary spermatocytes, cohesins at centromeres are cleaved by separase, leading to the separation of sister chromatids and to the production of four haploid round spermatids.

Before focusing on the process of spermiogenesis, it is of interest to note that energy metabolism in the testis shows some specificity in that neither spermatocytes nor spermatids metabolize glucose; instead, the lactate, supplied by Sertoli cells, is the main energy metabolite. In brief, glucose is transported into Sertoli cells via GLUT 1 and is processed in lactate. Lactate is then transferred to germ cells in the context of an active cooperation under control of endocrine and paracrine system (Boussouar and Benahmed 2004).

2.4.3 Spermiogenesis

Spermiogenesis represents the final stage of the spermatogenetic process in which the haploid non-flagellated round spermatids undergo a process of differentiation into elongated spermatozoa that are released into the seminiferous tubules (Fig. 2.3). During this stage, no further cell divisions take place and, instead, a profound reshaping of the cells with a number of cytoplasmic and nuclear alterations occur. The round spermatids are characterized by a small spherical nucleus; their cytoplasm is rich in the usual organelles, including mitochondria and Golgi apparatus, and the centriolar complex comprising the proximal and distal centrioles (Chen et al. 2017).

Here we will describe the main steps of spermiogenesis following the acrosomal and flagellum shaping and the condensation of chromatin, being aware that these processes occur in parallel.

The development of the acrosome of mammalian spermatozoa includes four different steps: Golgi phase, cap phase, acrosomal phase, and maturation phase (Muciaccia et al. 2013).

During the Golgi phase, the trans-Golgi network of the round spermatid, composed as usual of stacks of flattened membranous discs, produces proacrosomal granules and vesicles.

In the cap phase, the proacrosomal vesicles and other vesicles derived from Golgi apparatus and from endocytic pathways (Pleuger et al. 2020) fuse to form a large acrosomal vesicle that flattens and enlarges over the nuclear membrane (Fig. 2.4).

This consistent vesicle formation, transport, nuclear adhesion, and acrosome extension depend upon microfilaments, microtubules and related motor proteins (Pleuger et al. 2020).

A progressive elongation of the nucleus characterizes the acrosome phase; the acrosome follows the shape of the nucleus (Fig. 2.5a) covering, at the end of this process, the 2/3 of the nuclear surface (Fig. 2.5b). During maturation phase, whereas the nucleus continues to condense, the acrosome develops its species-specific shape and size.

The nuclear changes are not only restricted to the shape but also the chromatin is deeply remodeled. In fact, early germ cells present the typical chromatin

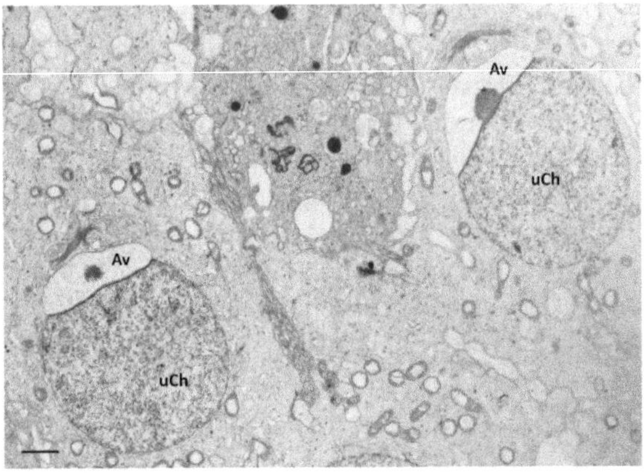

Fig. 2.4 Transmission electron microscopy micrograph of human testis seminiferous tubule showing round spermatids. The acrosomal vesicle is evident (Av); the round-shaped nuclei show granular uncondensed chromatin (uCh). Bar: 1 μm

organization of somatic cells in which DNA interacts with histones, positively charged proteins involved also in the mechanism of regulation of gene expression. One of the most evident characteristics of spermiogenesis is the phenomenon of chromatin condensation (Fig. 2.5a, b) and the replacement of histones by protamines, through a process involving transition proteins in several species including humans (Ribas-Maynou et al. 2022).

Protamines, which replace 85% of the nuclear histone content, contain high percentages of arginine and lysine, positively charged amino acids, and they can efficiently bind to DNA. In humans, protamine 1 and protamine 2 are present. The interaction between protamines and DNA leads to a peculiar, supercoiled structure called toroidal DNA–protamine complexes that are connected with the nuclear matrix. This high DNA condensation, that makes the sperm in most part transcriptionally silent, is important in the protection of genetic content delivered to the embryo.

Simultaneously with acrosome biogenesis, but at the opposite side of the spermatid elongating nucleus, the flagellum starts their shaping (Fig. 2.3).

The axonemal structure, composed by nine doublet microtubules and a central couple (9 + 2 pattern), emerges from the centrosome, a hub including two centrioles made up of nine triplet microtubules surrounded by a matrix of proteins called pericentriolar material (Tarozzi et al. 2021).

At the end of meiosis, the round spermatid inherits two canonical centrioles; the distal centriole (DC) docks to the cell membrane and starts to nucleate the axoneme, the bearing structure of the flagellum. The proximal centriole (PC) is located perpendicularly to the DC and develops a sort of "mini-flagellum" called "centriolar adjunct," probably involved in the development of the neck region. In the late stages of sperm maturation, the centriolar adjunct decreases in length and significantly reduces or disappears in human mature sperm.

When the flagellum starts growing, the centrioles are pushed toward the base of the elongating nucleus. In the mature spermatozoon, the PC that maintains the canonical barrel-shaped structure is located at the base of the head. The DC

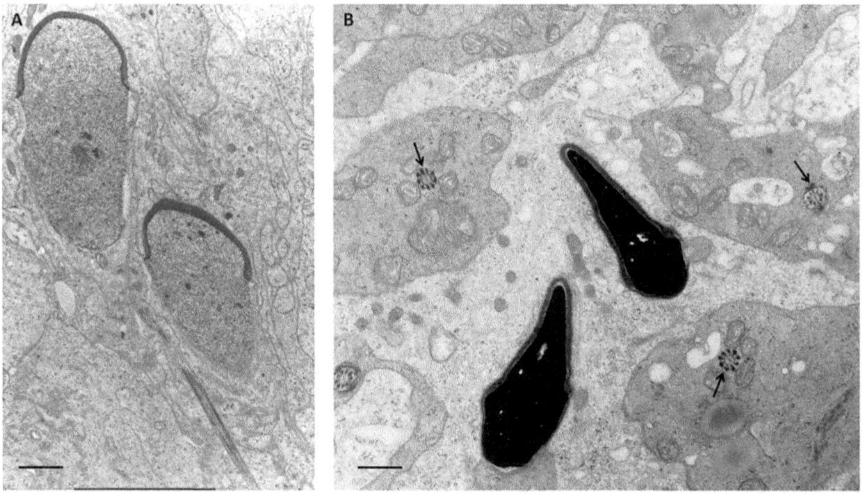

Fig. 2.5 (**A**, **B**) Transmission electron microscopy micrographs of human testis seminiferous tubule. In (**A**), elongating spermatids in the acrosomal phase. In (**B**), maturation phase spermatids are shown. The nuclei show regular shape with condensed chromatin and normal acrosome. Cross sections of flagella are embedded in large cytoplasmic residues (arrows). Bars: (**A**, **B**) 1 μm

disappears "based on centrosomal reduction model" (Manandhar et al. 2000) or is deeply modified based on "centriole remodeling hypothesis" showing a structure of nine splayed doublets of microtubules (Avidor-Reiss et al. 2019; Avidor-Reiss et al. 2020; Khanal et al. 2021); in any case, it organizes the axoneme.

It is recognized that the sperm tail includes also other three unique accessory structures: mitochondria around the midpiece, outer dense fibers (ODFs) and the fibrous sheath around axoneme. Today, the mechanisms by which the proteins constituting these structures are targeted to the destination are largely unknown, even though cytoskeletal transport is involved (Pleuger et al. 2020).

The mitochondria fold around the tail basis, shaping the midpiece. The majority of the cytoplasm is removed from the spermatids to maximize sperm motility and it is phagocytized by Sertoli cells (Neto et al. 2016). Mature spermatozoa are released into the lumen of the seminiferous tubules by means of a process called spermiation that is not fully understood in humans but seems to be under control of FSH and testosterone.

It is clear that during spermiogenesis, a crucial role is played by the cytoskeleton, and, in particular, by two exclusive structures, the acroplaxome and the manchette (Fig. 2.6), that collaborate with different proteins in acrosome biogenesis, nuclear elongation and flagellum formation.

Acroplaxome is a plate-like structure that anchors the acrosome to the nucleus (Fig. 2.6); microfilaments, motor proteins, keratin and other proteins are components of this structure. Acroplaxome is one of the essential cytoskeletal structures needed for the acrosome shaping.

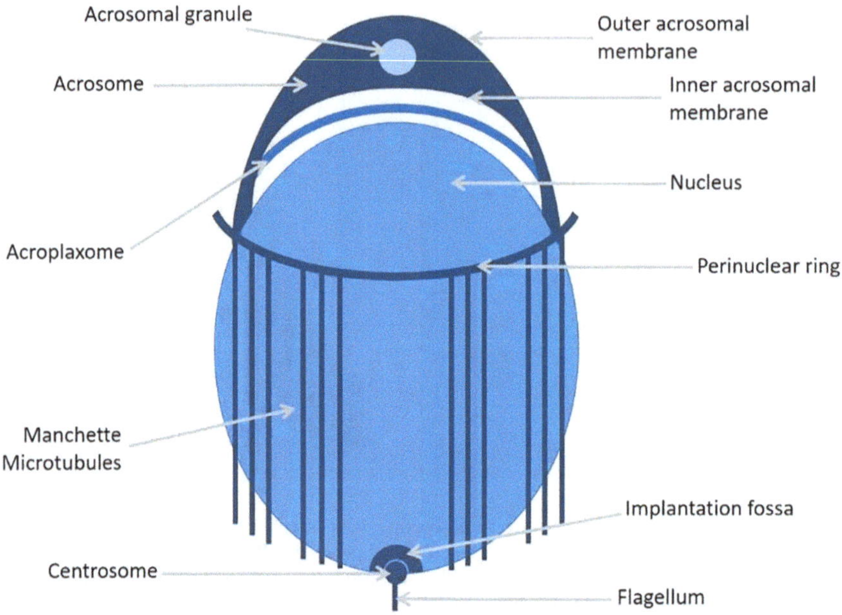

Fig. 2.6 Simplified representation of developing spermatids with elliptical nucleus surrounded by the perinuclear ring and by the manchette a peculiar microtubular structure. The acrosome, delimitated by outer acrosomal membrane and inner acrosomal membrane, is located in the apical part and attaches to the nucleus by the acroplaxome

The manchette, a cytoskeletal structure present only during spermatid elongation, is a transitory skirt-like structure based on microtubules surrounding the elongating spermatid nucleus (Wei and Yang 2018). The manchette plays a role in the transport of vesicles and macromolecules to the centrosome and developing spermatid tails as well as in nucleocytoplasmic transport (Lehti and Sironen 2016). In addition, Manfrevola et al. (2021) revised a peculiar structure that links the nucleoskeleton and the cytoskeleton, named LINC, involved in several functions needed for the correct development of sperm cells such as head formation and head to tail connection.

2.5 Conclusions

Spermatogenesis and spermiogenesis play a very important role in the constant production of numerous spermatozoa. The population of spermatogonial stem cells must be able to self-renew to keep the spermatogenetic process active; the maturation processes involve a series of significant morphological modifications that lead the early spermatid, a round haploid cell, to assume the typical shape of the spermatozoon. The knowledge of these processes is necessary because alterations in the intricate cross-talk occurring between germ cells and somatic cells, such as Sertoli cells, Leydig cells and peritubular myoid cells, can reduce the quantity and quality of spermatozoa undermining the fertilizing capacity of the subject.

References

Adelman CA, Petrini JH (2008) ZIP4H (TEX11) deficiency in the mouse impairs meiotic double strand break repair and the regulation of crossing over. PLoS Genet 3:e1000042. https://doi.org/10.1371/journal.pgen.1000042

Amann RP (2008) The cycle of the seminiferous epithelium in humans: a need to revisit? J Androl 29(5):469–487. https://doi.org/10.2164/jandrol.107.004655

Avidor-Reiss T, Mazur M, Fishman EL, Sindhwani P (2019) The role of sperm centrioles in human reproduction – the known and the unknown. Front Cell Dev Biol 7:188

Avidor-Reiss T, Carr A, Fishman EL (2020) The sperm centrioles. Mol Cell Endocrinol 518:110987. https://doi.org/10.1016/j.mce.2020.110987

Baazm M, Mashayekhi FJ, Babaie S, Bayat P, Beyer C, Zendedel A (2017) Effects of different Sertoli cell types on the maintenance of adult spermatogonial stem cells in vitro. In Vitro Cell Dev Biol Anim 53(8):752–758. https://doi.org/10.1007/s11626-017-0172-z

Bay K, Hartung S, Ivell R, Schumacher M, Jürgensen D, Jorgensen N, Holm M, Skakkebaek NE, Andersson AM (2005) Insulin-like factor 3 serum levels in 135 normal men and 85 men with testicular disorders: relationship to the luteinizing hormone-testosterone axis. J Clin Endocrinol Metab 90(6):3410–3418. https://doi.org/10.1210/jc.2004-2257

Birkhead TR, Montgomerie R (2009) Three centuries of sperm research. In: Birkhead TR, Hosken DJ, Pitnick SS (eds) Sperm biology: an evolutionary perspective. Elsevier, London, pp 1–42

Boitani C, Di Persio S, Esposito V, Vicini E (2016) Spermatogonial cells: mouse, monkey and man comparison. Semin Cell Dev Biol 59:79–88. https://doi.org/10.1016/j.semcdb.2016.03.002

Boussouar F, Benahmed M (2004) Lactate and energy metabolism in male germ cells. Trends Endocrinol Metab 15(7):345–350. https://doi.org/10.1016/j.tem.2004.07.003

Cannarella R, Condorelli RA, Mongioì LM, La Vignera S, Calogero AE (2020) Molecular biology of spermatogenesis: novel targets of apparently idiopathic male infertility. Int J Mol Sci 21(5):1728. https://doi.org/10.3390/ijms21051728

Carreau S, Bouraima-Lelong H, Delalande C (2012) Role of estrogens in spermatogenesis. Front Biosci (Elite Ed) 4:1–11

Chen LY, Willis WD, Eddy EM (2016) Targeting the Gdnf gene in peritubular myoid cells disrupts undifferentiated spermatogonial cell development. Proc Natl Acad Sci USA 113:1829–1834

Chen H, Mruk D, Xiao X, Cheng CY (2017) Human spermatogenesis and its regulation. In: Winters S, Huhtaniemi I (eds) Male hypogonadism. Contemporary endocrinology. Humana Press, Cham. https://doi.org/10.1007/978-3-319-53298-1_3

Cheng CY, Wong EW, Yan HH, Mruk DD (2010) Regulation of spermatogenesis in the microenvironment of the seminiferous epithelium: new insights and advances. Mol Cell Endocrinol 315(1–2):49–56. https://doi.org/10.1016/j.mce.2009.08.004

Clermont Y (1966) Renewal of spermatogonia in man. Am J Anat 118(2):509–524

Cooper A (1845) Observations on the structure and diseases of the testis. Lea and Blanchard, Philadelphia

Corona G, Rastrelli G, Vignozzi L, Maggi M (2022) Androgens and male sexual function. Best Pract Res Clin Endocrinol Metab 36(4):101615. https://doi.org/10.1016/j.beem.2022.101615

Edelsztein NY, Valeri C, Lovaisa MM, Schteingart HF, Rey RA (2022) AMH regulation by steroids in the mammalian testis: underlying mechanisms and clinical implications. Front Endocrinol (Lausanne) 13:906381. https://doi.org/10.3389/fendo.2022.906381

Facondo P, Delbarba A, Maffezzoni F, Cappelli C, Ferlin A (2020) INSL3: a marker of Leydig cell function and testis-bone-skeletal muscle network. Protein Pept Lett 27(12):1246–1252. https://doi.org/10.2174/0929866527666200925105739

Fayomi AP, Orwig KE (2018) Spermatogonial stem cells and spermatogenesis in mice, monkeys and men. Stem Cell Res 29:207–214. https://doi.org/10.1016/j.scr.2018.04.009

Franca LR, Hess RA, Dufour JM, Hofmann MC, Griswold MD (2016) The Sertoli cell: one hundred fifty years of beauty and plasticity. Andrology 4(2):189–212

French FS, Ritzen EM (1973) A high-affinity androgen-binding protein (ABP) in rat testis: evidence for secretion into efferent duct fluid and absorption by epididymis. Endocrinology 93(1):88–95

Geyer CB (2018) Publisher's note: a historical perspective on some "new" discoveries on spermatogenesis from the laboratory of Enrico Sertoli in 1878. Biol Reprod 99(3):671. https://doi.org/10.1093/biolre/ioy146

Grelon M (2016) Meiotic recombination mechanisms. C R Biol 339(7–8):247–251. https://doi.org/10.1016/j.crvi.2016.04.003

Griswold MD (2018) 50 years of spermatogenesis: Sertoli cells and their interactions with germ cells. Biol Reprod 99(1):87–100. https://doi.org/10.1093/biolre/ioy027

Heller CG, Clermont Y (1963) Spermatogenesis in man: an estimate of its duration. Science 140(3563):184–186. https://doi.org/10.1126/science.140.3563.184

Hermo L, Pelletier RM, Cyr DG, Smith CE (2010) Surfing the wave, cycle, life history, and genes/proteins expressed by testicular germ cells. Part 4: intercellular bridges, mitochondria, nuclear envelope, apoptosis, ubiquitination, membrane/voltage-gated channels, methylation/acetylation, and transcription factors. Microsc Res Tech 73(4):364–408. https://doi.org/10.1002/jemt.20785

Jarow JP, Chen H, Rosner TW, Trentacoste S, Zirkin BR (2001) Assessment of the androgen environment within the human testis: minimally invasive method to obtain intratesticular fluid. J Androl 22(4):640–645

Khanal S, Leung MR, Royfman A, Fishman EL, Saltzman B, Bloomfield-Gadêlha H, Zeev-Ben-Mordehai T, Avidor-Reiss T (2021) A dynamic basal complex modulates mammalian sperm movement. Nat Commun 12(1):3808

Lambard S, Galeraud-Denis I, Bouraïma H, Bourguiba S, Chocat A, Carreau S (2003) Expression of aromatase in human ejaculated spermatozoa: a putative marker of motility. Mol Hum Reprod 9:117–124. https://doi.org/10.1093/molehr/gag020

Lambard S, Silandre D, Delalande C, Denis-Galeraud I, Bourguiba S, Carreau S (2005) Aromatase in testis: expression and role in male reproduction. J Steroid Biochem Mol Biol 95:63–69. https://doi.org/10.1016/j.jsbmb.2005.04.020

Lehti MS, Sironen A (2016) Formation and function of the manchette and flagellum during spermatogenesis. Reproduction 151(4):R43–R54. https://doi.org/10.1530/REP-15-0310

Leslie SW, Mejias SG, Ramphul K (2023) Sertoli cell–only syndrome. In: StatPearls [internet]. StatPearls publishing, Treasure Island (FL)

Leydig DF (1850) Zur Anatomie der Mannlichen Geschlechtsorgane und Analdrusen der Saugethiere. Z Wiss Zool 2:1–57

Lucas-Herald AK, Mitchell RT (2022) Testicular Sertoli cell hormones in differences in sex development. Front Endocrinol (Lausanne) 13:919670. https://doi.org/10.3389/fendo.2022.919670

Maekawa M, Kamimura K, Nagano T (1996) Peritubular myoid cells in the testis: their structure and function. Arch Histol Cytol 59(1):1–13. https://doi.org/10.1679/aohc.59.1

Manandhar G, Simerly C, Schatten G (2000) Highly degenerated distal centrioles in rhesus and human spermatozoa. Hum Reprod 15:256–263. https://doi.org/10.1093/humrep/15.2.256[

Manfrevola F, Guillou F, Fasano S, Pierantoni R, Chianese R (2021) LINCking the nuclear envelope to sperm architecture. Genes (Basel) 12(5):658. https://doi.org/10.3390/genes12050658

Mayerhofer A (2013) Human testicular peritubular cells: more than meets the eye. Reproduction 145(5):R107–R116. https://doi.org/10.1530/REP-12-0497

Mendis-Handagama SM, Ariyaratne HB (2001) Differentiation of the adult Leydig cell population in the postnatal testis. Biol Reprod 65:660–671

Misell LM, Holochwost D, Boban D, Santi N, Shefi S, Hellerstein MK, Turek PJ (2006) A stable isotope-mass spectrometric method for measuring human spermatogenesis kinetics in vivo. J Urol 175(1):242–246; discussion 246. https://doi.org/10.1016/S0022-5347(05)00053-4

Mruk DD, Cheng CY (2015) The mammalian blood-testis barrier: its biology and regulation. Endocr Rev 36(5):564–591. https://doi.org/10.1210/er.2014-1101

Muciaccia B, Boitani C, Berloco BP, Nudo F, Spadetta G, Stefanini M, de Rooij DG, Vicini E (2013) Novel stage classification of human spermatogenesis based on acrosome development. Biol Reprod 89(3):60. https://doi.org/10.1095/biolreprod.113.111682

Nakagawa A, Shiratsuchi A, Tsuda K, Nakanishi Y (2005) In vivo analysis of phagocytosis of apoptotic cells by testicular Sertoli cells. Mol Reprod Dev 71(2):166–177. https://doi.org/10.1002/mrd.20278

Neto FT, Bach PV, Najari BB, Li PS, Goldstein M (2016) Spermatogenesis in humans and its affecting factors. Semin Cell Dev Biol 59:10–26. https://doi.org/10.1016/j.semcdb.2016.04.009

O'Donnell L, Meachem SJ, Stanton PG, McLachlan RI (2006) Endocrine regulation of spermatogenesis. In: Neill JD (ed) Physiology of reproduction, 3rd edn. Elsevier, Amsterdam, pp 1017–1069

O'Shaughnessy PJ (2014) Hormonal control of germ cell development and spermatogenesis. Semin Cell Dev Biol 29:55–65

O'Connor AE, De Kretser DM (2004) Inhibins in normal male physiology. Semin Reprod Med 22(3):177–185. https://doi.org/10.1055/s-2004-831893

O'Donnell L, Smith LB, Rebourcet D (2022) Sertoli cells as key drivers of testis function. Semin Cell Dev Biol 121:2–9. https://doi.org/10.1016/j.semcdb.2021.06.016

Oduwole OO, Huhtaniemi IT, Misrahi M (2021) The roles of luteinizing hormone, follicle-stimulating hormone and testosterone in spermatogenesis and Folliculogenesis revisited. Int J Mol Sci 22(23):12735. https://doi.org/10.3390/ijms222312735

Parekh PA, Garcia TX, Hofmann MC (2019) Regulation of GDNF expression in Sertoli cells. Reproduction 157(3):R95–R107. https://doi.org/10.1530/REP-18-0239

Planinić A, Marić T, Bojanac AK, Ježek D (2022) Reinke crystals: hallmarks of adult Leydig cells in humans. Andrology 10(6):1107–1120. https://doi.org/10.1111/andr.13201

Pleuger C, Lehti MS, Dunleavy JE, Fietz D, O'Bryan MK (2020) Haploid male germ cells-the grand Central Station of protein transport. Hum Reprod Update 26(4):474–500. https://doi.org/10.1093/humupd/dmaa004

Previato de Almeida L, Evatt JM, Chuong HH, Kurdzo EL, Eyster CA, Gladstone MN, Gómez-H L, Llano E, Meyer R, Pendas AM, Pezza RJ, Dawson DS (2019) Shugoshin protects centromere pairing and promotes segregation of nonexchange partner chromosomes in meiosis. Proc Natl Acad Sci USA 116(19):9417–9422. https://doi.org/10.1073/pnas.1902526116

Prince FP (2001) The triphasic nature of Leydig cell development in humans, and comments on nomenclature. J Endocrinol 168(2):213–216. https://doi.org/10.1677/joe.0.1680213

Ribas-Maynou J, Nguyen H, Wu H, Ward WS (2022) Functional aspects of sperm chromatin organization. Results Probl Cell Differ 70:295–311. https://doi.org/10.1007/978-3-031-06573-6_10

Sandlow JI, Feng HL, Cohen MB, Sandra A (1996) Expression of c-KIT and its ligand, stem cell factor, in normal and subfertile human testicular tissue. J Androl 17(4):403–408

Sertoli E (1865) Dell'esistenza di particolari cellule ramificate nei canalicoli seminiferi del testicolo umano. Il Morgagni 7:31–40

Shah W, Khan R, Shah B, Khan A, Dil S, Liu W, Wen J, Jiang X (2021) The molecular mechanism of sex hormones on Sertoli cell development and proliferation. Front Endocrinol (Lausanne) 12:648141. https://doi.org/10.3389/fendo.2021.648141

Sinclair AH, Berta P, Palmer MS, Hawkins JR, Griffiths BL, Smith MJ et al (1990) A gene from the human sex-determining region encodes a protein with homology to a conserved DNA-binding motif. Nature 346:240–244

Song J, Sha Y, Liu X, Zeng X, Zhao X (2023) Novel mutations of TEX11 are associated with non-obstructive azoospermia. Front Endocrinol (Lausanne) 14:1159723. https://doi.org/10.3389/fendo.2023.1159723

Stanton PG (2016) Regulation of the blood-testis barrier. Semin Cell Dev Biol 59:166–173. https://doi.org/10.1016/j.semcdb.2016.06.018

Sun Q, Liu F, Mo X, Yao B, Liu G, Chen S, Ren Y (2022) Shugoshin regulates Cohesin, kinetochore-microtubule attachments, and chromosomal instability. Cytogenet Genome Res 162(6):283–296. https://doi.org/10.1159/000528141

Tarozzi N, Nadalini M, Coticchio G, Zacà C, Lagalla C, Borini A (2021) The paternal toolbox for embryo development and health. Mol Hum Reprod 27(7):gaab042

Thumfart KM, Mansuy IM (2023) What are Sertoli cells? Historical, methodological, and functional aspects. Andrology 11(5):849–859. https://doi.org/10.1111/andr.13386

Verhoeven G, Hoeben E, De Gendt K (2000) Peritubular cell-Sertoli cell interactions: factors involved in PmodS activity. Andrologia 32(1):42–45

Walsh EL, Cuyler WK, McCullagh DR (1934) The physiologic maintenance of the male sex glands: the effect of androtin on hypophysectomized rats. Am J Physiol-Leg Cont 107(2):508–512

Watanabe Y (2005) Shugoshin: guardian spirit at the centromere. Curr Opin Cell Biol 17(6):590–595. https://doi.org/10.1016/j.ceb.2005.10.003

Wei YL, Yang WX (2018) The acroframosome-acroplaxome-manchette axis may function in sperm head shaping and male fertility. Gene 660:28–40. https://doi.org/10.1016/j.gene.2018.03.059

Welsh M, Saunders PTK, Atanassova N, Sharpe RM, Smith LB (2009) Androgen action via testicular peritubular myoid cells is essential for male fertility. FASEB J 23:4218–4230

Welter H, Huber A, Lauf S, Einwang D, Mayer C, Schwarzer JU, Köhn FM, Mayerhofer A (2014) Angiotensin II regulates testicular peritubular cell function via AT1 receptor: a specific situation in male infertility. Mol Cell Endocrinol 393(1–2):171–178. https://doi.org/10.1016/j.mce.2014.06.011

Xie C, Wang W, Tu C, Meng L, Lu G, Lin G, Lu LY, Tan YQ (2022) Meiotic recombination: insights into its mechanisms and its role in human reproduction with a special focus on non-obstructive azoospermia. Hum Reprod Update 28(6):763–797. https://doi.org/10.1093/humupd/dmac024

Zhou R, Wu J, Liu B, Jiang Y, Chen W, Li J, He Q, He Z (2019) The roles and mechanisms of Leydig cells and myoid cells in regulating spermatogenesis. Cell Mol Life Sci 76(14):2681–2695. https://doi.org/10.1007/s00018-019-03101-9

Zirkin BR, Papadopoulos V (2018) Leydig cells: formation, function, and regulation. Biol Reprod 99(1):101–111. https://doi.org/10.1093/biolre/ioy059

The Human Spermatozoon: Structure and Motility

3

Contents

Core Messages

The aim of this chapter is to dissect the human spermatozoon into its different parts and summarizes the current knowledge on the structure and function of the different components.

The spermatozoon is a sophisticated machine shaped to fertilize the egg and to participate in embryo development. The integrity of all the sperm components is necessary to accomplish the mysterious process of fertilization: the plasma membrane with its specific domains for oocyte-sperm interaction, the acrosome rich in enzymes for the penetration of the zona pellucida of the oocyte, the haploid nucleus with a characteristic shape and chromatin structure different from that of somatic cells, the mitochondria the "powerhouse" of the spermatozoon and the flagellum, the engine for the movement.

Abbreviations

AKAPs	A-kinase anchoring proteins
ALA	Alpha-linolenic acid

E. Moretti, G. Collodel, *The Human Spermatozoon*,
https://doi.org/10.1007/978-3-031-97868-5_3

ARA	Arachidonic acid
CABYR	Calcium-binding tyrosine phosphorylation-regulated protein
DC	Distal centriole
DHA	Docosahexaenoic acid
EPA	Eicosapentaenoic acid
FAs	Fatty acids
LA	Linoleic acid
ODF1	Outer dense fibre1
PAWP	Post-acrosomal sheath WW domain-binding protein
PC	Proximal centriole
(PLC)ζ	Phospholipase C
PPs	Protein phosphatases
PUFAs	Polyunsaturated FAs
ROS	Reactive oxygen species
SPATA6	Spermatogenesis associated 6
sRNA	Small RNA

3.1 The Sperm Membrane

The human spermatozoon is approximately 60 μm long and consists of a head and a flagellum, 5 μm and 55 μm long, respectively (Baccetti 1984; Fawcett 1975). Like any other cell, the spermatozoon is enveloped by a plasma membrane composed of a phospholipid bilayer interspersed with cholesterol, which is crucial for regulating membrane fluidity, and various proteins.

In the external layer, glycoproteins and glycolipids form the glycocalyx that protects sperm during their journey in male and female reproductive tracts and plays a role in regulating sperm function (Schröter et al. 1999). However, spermatozoa are polarized cells with distinct regions such as acrosome, equatorial segment, post-acrosomal sheath, midpiece and flagellum, and the plasma membrane composition reflects the particular function of the different domains. It is clear that the relative presence of lipids and lipid class components plays a crucial role in defining the properties of the plasma membrane and its optimal function (Garolla et al. 2018). Among the different lipids of the human sperm plasma membrane, the phospholipids are the most common, and of these, phosphatidylcholine and phosphatidylethanolamine represent the major components. The key constituents of phospholipids are the fatty acids (FAs) classified, according to the presence of double bonds in their chain structure, as saturated FAs with no double bonds, as monounsaturated fatty acids when a single double bond is present, and as polyunsaturated FAs (PUFAs) when having two or more double bonds (Zalata et al. 1998). The analysis of the FA pattern of membrane phospholipids has shown that the percentage of saturated FAs is around 50% (palmitic and stearic acids being the most prevalent), oleic acid is the principal monounsaturated FA and the total percentage of PUFAs is around 30% (Lenzi et al. 1996). Three families of PUFAs have been classified according to the distance of the first double bond to the methyl terminal of the hydrocarbon chain, i.e., n-3, n-6, and n-9.

Most of the sperm membrane characteristics depend on its fluidity, flexibility, fusion, and curvature, and these properties are dependent on the FA composition of phospholipids. Phospholipids with a high amount of PUFA increase the biomembrane fluidity and flexibility because of the presence of numerous double bonds, whereas saturated or trans-FAs would make it rigid.

Long-chain PUFAs in cell membrane phospholipids derive from the metabolism, including elongation and desaturation processes, of alpha-linolenic acid (ALA) an omega 3 PUFA and linoleic acid (LA) an omega 6 PUFA. They are essential FAs because they cannot be synthesized by animals. In particular, ALA produces eicosapentaenoic acid (EPA) and docosahexaenoic acid (DHA), whereas LA is converted into arachidonic acid (ARA) (Collodel et al. 2020). LA, ALA and their metabolites influence the reproductive function and fertility. Recently, Lass and Belluzzi (2019) observed that the total amount of omega 3 PUFAs in subjects with normal semen parameters was increased with respect to that of individuals with abnormal semen parameters. It was reported that, in normozoospermic subjects, four specific FAs such as palmitic, stearic, oleic acids, and DHA represented 74% of the total FA mass and the single amount of DHA corresponded to 43% of the total PUFA content (Zerbinati et al. 2016). In human sperm cells, DHA and palmitic acid represent the major PUFA and saturated fatty acid, respectively (Esmaeili et al. 2015).

In general, FAs needed for sperm membrane composition and function can derive from de novo synthesis or from diet; in particular, dietary sources of ALA, DHA, and EPA are crucial to maintain a sufficient supply in n-3 PUFA metabolism (Collodel et al. 2020). On the other hand, PUFAs represent one of the main targets of free radical attacks. For this reason, the sperm membrane is a biostructure particularly vulnerable and exposed to lipid peroxidation. The induced damages may lead to motility loss, an increase of sperm anomalies (Aitken et al. 2016) that affect sperm physiology and fertilization.

The definitive membrane lipid composition of ejaculated spermatozoa is accomplished only after epididymal maturation. During epididymal transit, an exchange of phospholipids between sperm plasma membrane and the epididymal fluids is evident (Garolla et al. 2018; Paul et al. 2021).

In addition, during the capacitation process, the plasma membrane of the sperm head undergoes reorganization that comprises the loss of cholesterol, the formation of lipid microdomains and the relocation of membrane molecules over its surface (Mortimer 2018). These facts suggest that the dynamic changes in sperm membranes play an important role in sperm function.

3.2 The Sperm Head

The typical human sperm head has a classic oval, flattened pear shape and comprises mainly two elements that are essential in the fertilization process: the acrosome and the nucleus (Fig. 3.1).

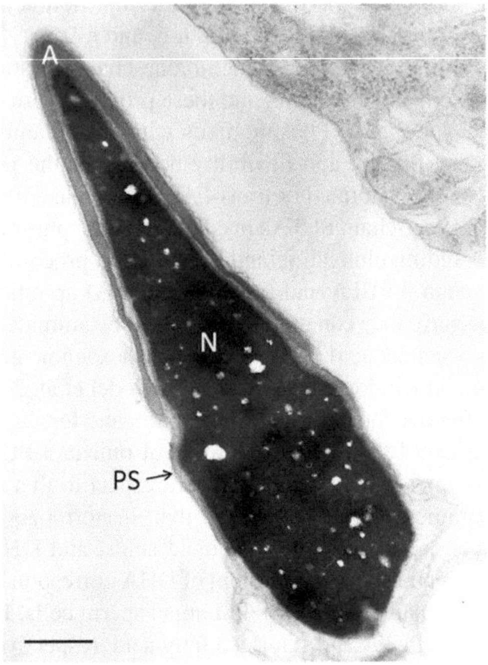

Fig. 3.1 Transmission electron microscopy micrograph of a longitudinal section of a head of normal human spermatozoon. *N* nucleus, *A* acrosome, *PS* post-acrosomal sheath. Bar: 0.5 μM

3.2.1 The Acrosome and the Post-acrosomal Region

The sperm acrosome develops by the interaction between early round spermatid nuclei and the adjacent Golgi complex. Golgi-derived vesicles contain hydrolytic enzymes that migrate, incorporate into the main acrosomal vesicle, and contribute to form the dense acrosomal granule. The acrosomal vesicle and granule spread on the anterior nuclear area to cover 75% of its surface (Fig. 3.1).

At the end of the shaping, the acrosome can be divided into a cap-like large anterior part, deputed to undergo acrosome reaction, and a posterior part called the equatorial segment that does not participate in the acrosome reaction; rather, the latter represents the area where the fusion between sperm and egg starts during fertilization (Mortimer 2018).

The acrosome biogenesis is closely associated with nuclear formation. Acrosome malformation is often associated with nuclear deformity, which is observed in human teratozoospermic sperm and in spermatozoa of gene-knockout infertile mice (Wei and Yang 2018).

In the region where the acrosome is located, the spermatozoon shows layers of membranes: the plasma membrane, the outer and inner acrosomal membranes, the nuclear envelope (Ito and Toshimori 2016). During its biogenesis, the acrosome collects different molecules needed for fertilization since it is involved in acrosome reaction, binding of and penetration through the zona pellucida of the oocyte, and sperm–egg membrane fusion. The acrosome reaction consists in the fusion of the outer acrosomal membrane and the plasma membrane, producing vesicles that release the acrosomal contents, in particular lithic enzymes with high proteolytic activity, such as hyaluronidase, esterase, peptidase, phospholipase, and acrosine (Zaneveld and De Jonge 1991).

However, the acrosome contains numerous proteins that are released and exposed during acrosomal reaction in an orderly manner depending on their localization and their role in the fertilization process. Here, we briefly present only a few of the relevant acrosomal proteins (for exhaustive presentation of these acrosomal proteins see Ito and Toshimori 2016).

Acrosin is a serine protease located in the acrosomal matrix and is initially synthesized as preproacrosin, which is then converted into proacrosin. The proacrosin is located in the apical segment and in the principal segment of the anterior acrosome (Baba et al. 1989).

Although the role of acrosin has been quite debated in these few decades, recently Hirose et al. (2020) demonstrated that acrosin is involved in penetration of zona pellucida as suggested by acrosin-knockout hamsters. The acrosome, in fact, interacts with the cumulus oophore and with the zona pellucida of the oocyte allowing the penetration of the spermatozoon.

Equatorin is a membrane protein that is specifically localized to the acrosomal membrane (Toshimori et al. 1998). It is not essential for acrosome biogenesis but is involved in the fusion of the outer acrosomal membrane and the sperm plasma membrane during the acrosome reaction (Hao et al. 2014).

IZUMO1 is a candidate sperm protein for sperm–egg fusion; Inoue et al. (2005) created a gene-disrupted mouse line and these IZUMO−/− mice were healthy, but males were unable to fertilize. IZUMO is maintained inside the acrosome and is not visible on the surface of mature intact sperm. At a very early stage of the acrosome reaction, IZUMO becomes detectable in the acrosomal cap and then on the entire surface of the head, including the apical acrosome, the posterior part of the acrosome, and the post acrosomal region (Ito and Toshimori 2016). A testis-specific serine kinase (Tssk6) is involved in the relocation of Izumo1 through the regulation of actin polymerization after acrosomal reaction (Sosnik et al. 2009).

The interaction between spermatozoon and oocyte is achieved through a two-step mechanism in which the sperm protein IZUMO on the equatorial segment of the acrosome-reacted sperm recognizes its receptor, JUNO, on the egg surface (Aydin et al. 2016).

Recently, Gómez-Torres et al. (2022) reported that human spermatozoa require the presence of sperm adhesion molecule 1 -SPAM1-endowed with hyaluronidase activity, throughout the sperm head to properly contact the cumulus-oocyte matrix.

The equatorial segment is the region of the spermatozoon that marks the end of the acrosomal region and the origin of the post-acrosomal sheath. The equatorial segment plays a pivotal role in human fertilization; it remains intact after the acrosomal reaction, and underlies the area of the plasma membrane involved in fusion with the membrane of the oocyte. The equatorial segment represents the site where the sperm nuclear envelope starts to breakdown after fertilization (Wolkowicz et al. 2008).

The post-acrosomal sheath (Fig. 3.1) is a crucial area of the human spermatozoon during fertilization. It represents a wide belt-like structure enveloping the posterior area of the sperm head (Mortimer 2018). The post-acrosomal sheath is the region where the post-acrosomal sheath WW domain-binding protein (PAWP) and the testis-specific phospholipase C (PLC)ζ are located. These proteins seem to be released into the oocyte after the gamete fusion, and trigger a series of molecular events known as "oocyte activation" (Aarabi et al. 2014; Kashir et al. 2010, 2015). Between the two candidates for oocyte activation, PLC ζ seems to be the most likely, at least in vitro (Nomikos et al. 2015).

In any case, the activation of oocyte by the spermatozoon is under control of the Ca^{2+} ions that are released in an oscillatory manner from the endoplasmic reticulum and control oocyte meiotic resumption, pronuclear formation, and embryo cleavage.

Notably, a number of clinical reports have highlighted the crucial importance of PLCζ in human and mammalian fertilization (Gat and Orvieto 2017) by associating PLCζ deficiencies with certain forms of male factor infertility (Saleh et al. 2020).

3.2.2 Nucleus and Chromatin

The nucleus contains a haploid chromosomal set and highly condensed chromatin, which appears dark at electron microscopy level (Fig. 3.1). Its shape is oval with the conical upper part and convex posterior area in which the flagellum is implanted.

At light and electron microscopy level, normal nucleus can show small vacuoles, considered to be physiological structures, large vacuoles are generally considered abnormal. The nature and origin of the nuclear vacuoles remain to be elucidated even though, recently, Gómez-Torres et al. (2023) proposed that they are cell structures originating from nuclear invaginations and containing perinuclear theca components.

The nuclear sperm membrane shows unique characteristics related to the fact that the mature spermatozoon is inactive and reputed to be incapable of transcription, even though it was demonstrated that sperm condensed chromatin can be partially transcribed (Santiago et al. 2021). In fact, the nuclear envelope lacks nuclear pores except in a small area corresponding to the site of head-tail attachment (Mortimer 2018); the persistent presence of nuclear pores in this region suggests that these structures can play a role in late stages of sperm maturation or during fertilization process (Ho 2010).

One of the most important characteristics of sperm nucleus is the high level of chromatin compaction. In early germ cells, from spermatogonia to round spermatids, which represent the first haploid cell, the DNA shows the typical organization found in somatic cells and histones are responsible for the first levels of chromatin packaging (the nucleosome formation). Histones are proteins characterized by a high percentage of positively charged amino acids and, other than chromatin condensation, they are involved in the regulation of gene expression (Alberts 2015).

Then, the chromatin interacting with other proteins, with nuclear matrix and nuclear lamina undergoes higher levels of organization. This configuration concerns the germ cells until they reach the haploid stage, which is the early spermatid stage. During spermiogenesis, the chromatin remodeling process occurs in a stepwise fashion. The canonical histones are removed, replaced by testis-specific histones (Tachiwana et al. 2008) transition proteins 1 and 2 (Zhao et al. 2004) and finally by protamines at later stages of spermatogenesis, resulting in very stable and highly compacted DNA (de la Iglesia et al. 2023; Moritz and Hammoud 2022; Ward 2010). This process is still quite obscure. One of the first histone variants discovered belongs to the H2B family and was named TH2B; it is characterized by an increased instability compared to a canonical octamer; then many other histone variants were found. Briefly, a coordinate action of histone variants opens the nucleosomes, allowing the loading of transition proteins which in turn orchestrate stepwise histone replacement and protamine assembly (Hoghoughi et al. 2018).

Protamines are composed by 50–110 amino acids and are much smaller than histones. Because of their increased lysine and arginine content, they also possess a stronger positive charge, which enables them to bind tightly to the main groove of DNA. Two families of protamines are present, in mammalian genome: protamine 1 and protamine 2. Protamine 1 is present in all mammalian spermatozoa and is important for correct chromatin condensation, while protamine 2 was observed in some species including humans (Corzett et al. 2002; Oliva 2006). Protamine 2 seems essential in increasing condensation operated by protamine 1; a hyper-expression of protamine 1 can cause alteration of sperm morphology and early DNA condensation (Ribas-Maynou et al. 2022).

The binding of DNA to protamines leads to the formation of toroidal, donuts-like, DNA-protamine complexes. The characteristics of the protamines and the high content of cysteine induce the formation of disulphide bonds between nearby protamines, stabilizing the toroidals structures. However, a part of sperm DNA remains associated to modified histones that might have a role in early embryo development (Ribas-Maynou et al. 2022). Although the sperm DNA is highly condensed by protamines into toroids, it seems that it is also organized into loop-domains interacting with nuclear matrix proteins. Ribas-Maynou et al. (2022) proposed a "toroid-loop model" for sperm chromatin organization, which includes different aspects of sperm DNA packaging observed by many scientists. In particular, the model suggests that every DNA loop domain in the sperm cells is coiled around a single protamine toroid. Sperm maturation continues through the male reproductive tract, during epididymal storage and transit, and biochemical changes should occur in the spermatozoa to become fertilizing competent. One of the most important changes is the active formation of disulfide bridges between thiol groups of cysteine residues present in protamines that make the sperm chromatin more compacted. The nuclear form of glutathione peroxidase 4 seems involved in sperm DNA condensation during epididymal maturation (O'Flaherty 2019).

The spatial macromolecular architecture of the chromatin protects and maintains the integrity of DNA during the sperm journey toward the oocyte, making it resistant to genotoxic agents and to oxidative insult, and preventing alterations of genetic information.

However, after fertilization, this highly compacted chromatin must be restored to its original state as a functionally active chromatin. To accomplish this, protamines must be replaced by histones in the male pronucleus, a process that is also hypothesized to involve the import of additional proteins into the paternal genome (Ribas-Maynou et al. 2022); in this way, the paternal chromatin has the same accessible chromatin as all other somatic cells.

The existence of this unique sperm chromatin packaging has important consequences for both the development of improved diagnostics for medical infertility, and for the study of higher order DNA structures in the field of cell biology.

Covering the topic concerning sperm nucleus, we should briefly mention the important epigenetic processes occurring in spermatozoa playing a relevant role in transmitting information from father to offspring (Bohacek and Rassoulzadegan 2020). Small RNA (sRNA) population, post-translational alteration of histones that control chromatin function and DNA cytosine methylation are the three most carefully studied epigenetic information carriers. In particular, different classes of sRNAs have been discovered and reputed to be implicated in the intergenerational inheritance of paternal environmental effects: microRNAs, gamete—PIWI-interacting RNAs particularly relevant in germ cells (see Sect. 4.7), as well as relatively understudied sRNAs, such as fragments derived from rRNAs and tRNAs (Liu and Sharma 2023; McSwiggin and O'Doherty 2018).

Classically, spermatozoa were seen as "transcriptionally inactive" (but little transcription is present) vehicles for delivering the paternal DNA to an egg. Yet, it has become evident that sperm also carry thousands of different RNAs and the functions of most of them are unknown. The fact that sperm have until now been presumed to only contribute DNA (and centromeres) to eggs, makes the recent discovery of a complex sperm RNA payload both surprising and hard to explain.

3.3 The Sperm Connecting Piece and the Midpiece

The sperm connecting piece is an important region that ensures head-tail attachment. During spermiogenesis, the connecting piece organizes around a pair of centrioles located in the caudal pole of spermatid nucleus and proteins belonging to the pericentriolar materials (Chemes 2017; Tarozzi et al. 2021). At spermatid stage, both proximal centriole (PC) and distal centriole (DC) show the typical cylinder shape with the wall made by 9 longitudinal triplets of microtubules in a circular arrangement (Chemes 2017). During spermiogenesis, the connecting piece undergoes important reshaping, mentioned also in the chapter related to the spermatogenetic process (see Sect. 2.4.3).

Briefly, during sperm maturation, the PC maintains its typical barrel structure, but the DC undergoes a profound remodeling. The question of centrioles in human spermatozoa is still debated. In the classical hypothesis, the early sperm's centrioles and the surrounding pericentriolar material are reduced by a process identified as "centrosome reduction." In particular, the DC disintegrates and the proteins of pericentriolar materials are eliminated. By this process, a vault is visible at transmission electron microscopy level in the area where the DC is expected to be. Following the dogma of centrosomal reduction, no functional DC is present in human spermatozoa (Manandhar et al. 2000). Recently, this line of research was opened again and the group of Avidor-Reiss, using sophisticated techniques demonstrated that two centrioles (the PC and a deeply modified DC) are present in the spermatozoa of non-murine-mammals (Fishman et al. 2018); by these results the "centriole remodeling hypothesis" was proposed, even though not universally accepted yet.

Following the "centriole remodeling hypothesis," the DC is made by nine splayed doublet microtubules forming a novel atypical structure and containing a subset of centriolar proteins; among them, centrin 1 is one of the proteins required for maturation of the DC (Avidor-Reiss and Fishman 2019, 2020; Fishman et al. 2018; Moretti et al. 2023a). DC surrounds the vault present in the connecting piece between the PC and the beginning of the axoneme and forms a sort of inverted cone (Fishman et al. 2018). PC and DC seem to function as centrioles in the zygote, both able to act as platform for the formation of a daughter centriole and to form regular spindle poles. The PC and the modified DC are enclosed by 9 striated columns close on the top by the capitulum or capitellum (Chemes 2017; Shen et al. 2020) forming a sort of box around the vault that stabilizes sperm head-tail junction. The nine segmented columns are in continuity with the corresponding outer dense fibers (ODF), surrounding the axoneme and present in the midpiece and principal piece. The capitulum anchors the connecting piece to the caudal pole of the sperm head at the implantation fossa and is attached to the basal plate.

Defects of this elaborate and organized region are at the basis of specific sperm pathologies characterized by alteration in the correct head-tail alignment up to the separation of sperm head from the tail (Chemes 2017; Elkhatib et al. 2017; Mortimer 2018).

Several genes are reported to be involved in the correct assembly and function of the connecting piece (Beurois et al. 2020; Zhu et al. 2018). A sperm-specific protein, outer dense fiber 1 (ODF1) seems to be essential for tight connection of the sperm tail to the head. ODF1 under-expression causes male infertility with spermatozoa with reduced basal plate, absence or poor development of striated columns and thin dense fibers (Hetherington et al. 2016). Spermatogenesis associated 6 (SPATA6) is also absolutely relevant in the development of connecting piece (Yuan et al. 2015). SPATA6 is a protein expressed in elongating and elongated spermatids as well as in mature spermatozoa and is localized in the segmented columns and in the capitulum.

The hallmark of the midpiece (generally, 5–7 μm long) is the presence of mitochondria that extend from the connecting piece to the septin-based annulus (Fig. 3.2), the ring-shaped structure which marks the edge between midpiece and principal piece.

A limited number of elongated mitochondria, arranged in a helical pattern of 10–12 gyres, are present in mature human spermatozoon, because the vast majority of them are lost in the residual cytoplasm eliminated during spermiogenesis (Fawcett 1970). Normal spermatozoa contain mitochondria with few mitochondrial (mt) DNA copy number; the mtDNA amount, and the occurrence of large mtDNA deletions are correlated with sperm quality (Boguenet et al. 2021).

From an ultrastructural point of view, the mitochondrial helix surrounds, from the inner to the outer region, the axoneme (described in the section dedicated to the

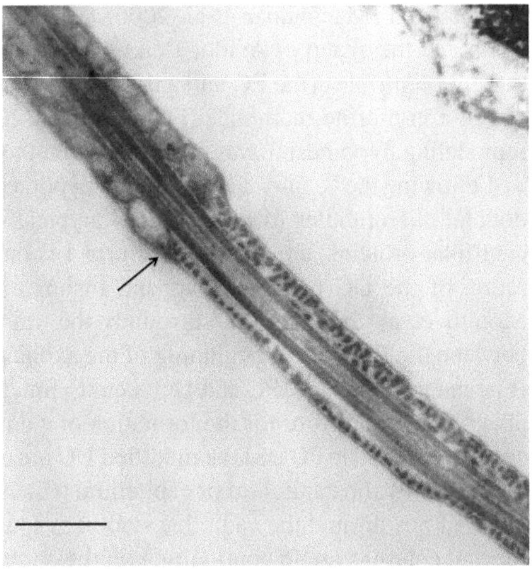

Fig. 3.2 Transmission electron microscopy micrograph of longitudinal section of a human sperm tail at level of the transition between midpiece and principal piece. The arrow indicates the annulus. Bar: 1 μm

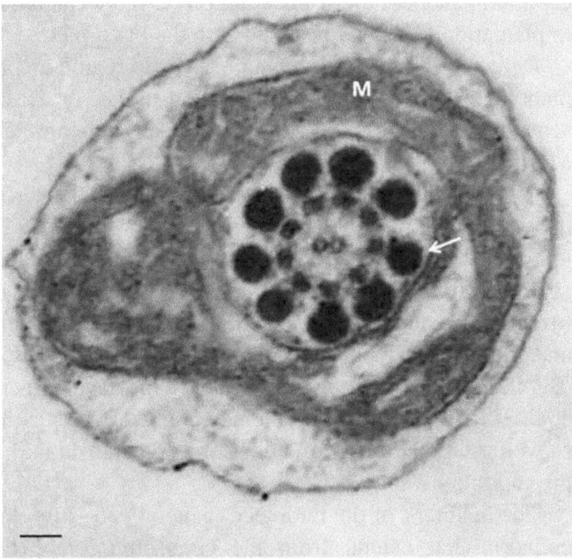

Fig. 3.3 Transmission electron microscopy micrograph of cross-section of the midpiece of a human spermatozoon. Mitochondria (M) surround the nine outer dense fibers (arrow) and the axoneme. Bar: 0.5 μm

principal piece), and the 9 ODFs, numbered from 1 to 9 (Fig. 3.3), which progress through the principal piece. In particular, in the midpiece, each ODF fiber is closely associated to the corresponding microtubular axonemal doublet; in the principal piece, ODFs 3 and 8 are not present.

More than 14 polypeptides are associated with ODFs including four major proteins, ODFs 1–4 whose gene expression is a post-meiotic event (Zhao et al. 2018).

The principal role of mitochondria is to provide ATP needed for motility, capacitation, hyperactivation, acrosome reaction, and fertilization (Du Plessis et al. 2015).

It seems that the ATP produced at mitochondrial level could be insufficient to diffuse to the entire flagellum and it is supported by ATP produced by glycolytic enzymes present in the fibrous sheath, a periaxonemal structure that characterized the principal piece (Gu et al. 2019). Overall spermatozoa are able to use different energy source from lactate to fatty acids depending on their availability in the surrounding environment (Boguenet et al. 2021) with AMP-activated protein kinase (AMPK) being a cellular energy sensor and a regulator of mitochondrial activity (Martin-Hidalgo et al. 2018).

Mitochondria represent the sites where oxidative phosphorylation and electron transport take place; because of the movement of electrons, they are involved in reactive oxygen species (ROS) production. Physiological ROS levels are essential to the activation of the pathways involved in normal sperm function (Aitken 2017; Moretti et al. 2023b) and are regulated by an antioxidant battery present in epididymis (O'Flaherty 2019), in the prostate, and seminal vesicle secretions. A ROS overproduction causes oxidative stress (OS) that induces a series of harmful processes in spermatozoa, among them, membrane lipoperoxidation, motility impairment, DNA damages (Aitken 2017; Drevet et al. 2022; Gibb et al. 2020; Moretti et al. 2023b).

The inner mitochondrial membrane is rich in peculiar proteins influenced by ROS. The prohibitin, a chaperone-like protein, is correlated with the production of ROS and involved in regulation of motility (Wang et al. 2012; Chai et al. 2017). Also, the mitochondrial uncoupling protein 2 (UCP2) is able to eliminate ROS, regulating sperm motility (Wang et al. 2018).

Mitochondria are also involved in the calcium ion homoeostasis, in the intrinsic apoptotic pathway, and steroid hormone biosynthesis (Amaral et al. 2013). Mitochondria of the oocyte are essential to the development of the embryo at early stages, while sperm mitochondria are important for sperm function and fertilization, but they seem to be non-essential after fertilization (Boguenet et al. 2021). In fact, in most organisms, the mitochondrial genome follows a uniparental maternal transmission, which means that the mitochondrial DNA of offspring is inherited exclusively from the mitochondria of the oocyte.

One of the reasons of this phenomenon could be that mitochondrial DNA is a preferential target of damages due to OS and its transmission to the embryo is not a favorable event. Many mechanisms are evolved to avoid the transmission of paternal mitochondrial DNA to the embryo. Among them, to be mentioned: the severe reduction of mitochondrial DNA copies during mammalian spermatogenesis (Amaral et al. 2007), the "simple dilution model" by which the copy number of paternal mitochondrial DNA is lower than that of maternal one (Yan et al. 2019), the destruction of paternal DNA or mitochondria during early embryo stages. We still do not know the exact mechanism involved in humans.

3.4 The Sperm Principal Piece

Distal to the annulus is the principal piece that presents the central axoneme surrounded by periaxonemal structures as a set of ODFs and the fibrous sheath.

The axonemal structure consists of a ring of nine microtubular doublets with two microtubules located in the center: the typical 9 + 2 pattern. Each doublet is formed by two closely associated microtubules: the microtubule A formed by 13 protofilaments and the incomplete microtubule B containing ten protofilaments and sharing part of its wall with microtubule A. At the beginning of the studies on spermatozoa, amino acid composition of "tubulin" constituting microtubules of sperm axoneme was essential to clarify the mechanisms underlying flagellar and ciliary movement (Mohri 1968).

The motor protein is the dynein arranged as two arms, outer one and inner one, which are anchored to the microtubule A and directed to the microtubule B of their next doublet in a clockwise direction to a viewer looking from the flagellar base to the tip (Linck et al. 2016). Dynein arms are mechanochemical force-transducing enzymes with ATPase activity and produce the sliding force needed for axonemal bending and flagellar motility. Outer and inner arms play different roles in the production and regulation of flagellar motility. The bending of the flagellum is due to the subsequent cycles of dynein arm attachment, generation of forces and detachment on the B subunit of the adjacent microtubule doublet. The outer arm is involved in the elevation of microtubule sliding velocity (increasing beat frequency), whereas the inner arm is implicated in the formation and propagation of flagellar bending (Inaba 2011).

The protein complex called nexin acts as a link between adjacent microtubular doublets and as elastic element to contain the amount of sliding and to return the doublets to their resting position.

The radial spokes projects from each of the nine doublets toward the central pair of microtubules and play a role in regulating sperm motility. Finally, the two central microtubules are joined by a bridge and surrounded by an inner sheath (Linck et al. 2016) to form the central pair microtubule apparatus. To achieve oscillation of flagellar bending, the dyneins on one side must be active while those on the other side are inactive to bend the axoneme. The radial spoke/central pair apparatus plays an essential role in the regulation of flagellar bending (Inaba 2011).

Numerous models have been proposed for sperm movement; most of them reported that spermatozoa motility is based on swimming movement. Essentially, the human spermatozoa exhibit sperm motility by moving the tail symmetrically from side to side (Kumar and Singh 2021).

Recently, Gadêlha et al. (2020) used high-speed three-dimensional microscopy and suggested that spermatozoa do not exhibit a simple swimming motion, but they roll during swimming. In other words, to reach a progressive motility, the spermatozoa spin in a corkscrew manner and drill through the fluid (Hansen et al. 2018).

The 9 + 2 microtubule structure is regulated by many microtubule-associated proteins. Some of them form helical structures within the microtubule lumen that have been postulated to be involved in controlling the swimming direction of spermatozoa (Zabeo et al. 2018).

Two additional cytoskeletal structures organize around the axoneme, the nine ODFs (9 + 9 + 2 cross-section pattern) and the fibrous sheath. At midpiece level, each ODF fiber is located in correspondence of each microtubular doublet; however, ODFs 3 and 8 do not continue to the principal piece and are replaced by the fibrous sheath that forms two longitudinal columns. In the principal piece, the size of other ODFs gradually decreases toward the end piece where they disappear. Researches have shown that ODFs help to preserve the flagellar flexibility and protect the sperm tail from stresses occurring during ejaculation and epididymal transit (Lehti and Sironen 2017; Zhao et al. 2018). ODFs are characterized by a high level of disulphide cross-linking mainly acquired during epididymal transit and their structural thickness determines flagellar bending curvature and regulate motility (Kadam et al. 2020).

The fibrous sheath is a structure characteristic of the principal piece and assembles around the axoneme-ODF complex. It is made up by a pair of longitudinal columns, which occupy the place of ODF 3 and 8 present in the midpiece, connected by a double series of semicircular ribs (Eddy et al. 2003). The fibrous sheath influences the way the spermatozoa move in terms of plane of beating of flagellum, flexibility and the shape of the flagellar wave (Mortimer 2018).

The majority of the genes encoding the components of fibrous sheath are expressed during spermiogenesis, when the meiotic process is completed. Among the different categories of fibrous sheath proteins, several enzymes of glycolytic pathway are relevant and indicate that most of the ATP needed for sperm motility is generated by glycolysis. In fact, most of the glycolytic enzymes have been localized to the principal piece of sperm flagellum (Eddy 2007).

However, the most relevant proteins that are part of the fibrous sheath are included in the class of A-kinase anchoring proteins (AKAPs) characterized by anchoring sites for cAMP-dependent protein kinase (Miki et al. 2002). AKAP4, the most abundant one, AKAP3 that is a site of attach for ropporin, TAKAP-80 act as anchoring site for protein kinases. In particular, AKAP4 is produced from a precursor, the proAKAP4 which acts as a disposable stock of AKAP4 to ensure long-lasting sperm motility up to the oocyte fertilization. Therefore, proAKAP4 can represent a sort of stress sensor that has the ability to be converted into mature AKAP4 or degraded by proteolysis and, in this way, can function as a possible gatekeeper to prevent transmission of harmful genetic damage to the next generation (Carracedo et al. 2022). Recently, Yu et al. (2023) observed that serine/threonine kinase 33 (STK33) regulates the phosphorylation of AKAP3 and AKAP4 affecting the fibrous sheath assembly and playing a relevant role in spermiogenesis and male infertility.

Another important protein of the fibrous sheath is the calcium-binding tyrosine phosphorylation-regulated protein (CABYR) that is involved in capacitation. CABYR can form complex not only with the scaffolding protein AKAP3 but also with ropporin (Li et al. 2011), a binding protein for rhophilin that interacts with GTP-ase Rho involved in the regulation of sperm motility (Fujita et al. 2000).

The downregulation of both ropporin and CABYR observed in asthenozoospermic samples and the positive correlation between the expression of the two genes suggests that the co-expression of these two proteins is a prerequisite for sperm motility (Pelloni et al. 2018).

Finally, the end piece, characterized by the lack of ODFs and fibrous sheath, is thinner than the other parts of the flagellum. It contains only the axoneme, which progressively loses its characteristic structure, enveloped by the plasma membrane.

3.5 Brief Summary on Sperm Motility Activation

The goal of the spermatozoon is to fertilize the oocyte and to accomplish this purpose it must be motile. After spermiogenesis, the spermatozoa released in seminiferous tubules are immotile, they acquire motility throughout the epididymal transit and exhibit a hyperactivated motility near the oocyte in female reproductive tract.

First, the prerequisites for sperm motility are both structural integrity of sperm flagella and the energy availability. ATP is the energy-carrying molecule used by axonemal dynein ATPases within the flagellum and ATP is needed also for protein tyrosine phosphorylation, hallmark of sperm capacitation (Freitas et al. 2017). Sperm oxidative phosphorylation occurs in mitochondria, localized at the beginning of the flagellum in the midpiece; the ATP should then be efficiently transported to the whole length of the flagellum. In addition, other ATP molecules can also be produced all along the principal piece by the glycolytic process by which a glucose molecule is converted into pyruvate and the energy is released in the form of two ATP molecules and NADH (Ford 2006). Moreover, a sperm-specific lactate dehydrogenase is anchored all along the fibrous sheath and represents a local ATP production near the site of ATP consumption (Miki et al. 2004).

So, the midpiece and the principal piece are different in structure in functioning and contain different signaling molecules and ion channels involved in sperm motility activation.

As reported earlier, the first activation of motility appears during epididymal transit, in particular in the last segment of epididymis: the cauda. It is known that epididymal maturation requires the acquisition and loss of proteins synthesized and secreted in the lumen by the cells of epididymal epithelium in a region-dependent fashion with the involvement of the epididymosomes. Epididymosomes, released from the epididymal epithelium, contain proteins, noncoding RNAs and distinct set of lipids that are transferred to spermatozoa while they pass through the different epididymal regions (Paul et al. 2021). Despite the numerous new achievements on the way epididymis works, the exact mechanism behind motility acquisition is still unknown, in particular, in humans since the study on epididymis are mostly performed in rodent models due to the scarce availability of human epididymal tissue that can be collected by subjects in reproductive age (Freitas et al. 2017).

In the epididymis, one of the first observed signaling events leading to sperm motility acquisition is the regulation of specific isoforms of protein phosphatases (PPs). PP activities decrease during sperm epididymal maturation, whereas protein kinase activity increases these two major events are required for motility activation (Pelech et al. 2013; Ferreira et al. 2022). For example, PP1 activity is high in immotile spermatozoa from caput epididymis and is low in motile spermatozoa from cauda epididymis (Freitas et al. 2017). The current mechanisms based on PPs and its regulators have been extensively investigated, but are still far from being fully understood.

Other mechanisms involved in motility activation are based on calcium ion, which is particularly relevant in triggering hyperactivated motility, landmark of capacitation, in the female reproductive tract. In spermatozoa, the calcium homeostasis is regulated by two main Ca^{2+} channels: Orai1 channel that regulates store-operated calcium influx (Davis et al. 2016), and the CatSper channel, which is the most extensively studied Ca^{2+} channels in mammalian sperm (Sun et al. 2017), and controls the intracellular calcium concentration. CatSper channel is a protein complex containing four α subunits named Catsper1–4 that form the pore and other auxiliary subunits involved in several functions as modulation by external signal, compartmentalization of calcium signaling and others.

CatSper channels form functional domains within the principal piece to allow fast and easy calcium signal propagation along the flagellum. However, this channel is able to mediate a rapid calcium influx also in the midpiece and sperm head (Rahban and Nef 2020; Singh and Rajender 2015).

The relevant role of Catsper during fertilization is confirmed by the fact that null mice lacking any CatSper isoform are infertile as well as men with mutations in CatSper genes. The acquisition of hyperactivated motility is essential for enabling spermatozoa to detach from oviductal epithelium where they form a sort of reservoir, to cross the protective coat of oocyte.

CatSper can be activated in a ligand-dependent pathway by progesterone, prostaglandins, cyclic nucleotide cAMP, albumin, and zona pellucida glycoproteins and in a ligand independent pathway by intracellular alkaline pH, change in membrane voltage (Brenker et al. 2012; Rahban and Nef 2020).

To this end, many membrane channels are involved in CatSper and motility activation. As a pH-sensitive ion channel, H^+ is the major regulator of acid-base microenvironment influencing CatSper. Two main channels are involved in the alkalization of spermatozoon: Na^+/H^+ exchangers import Na^+ into the plasma membrane and export H^+ out of the spermatozoa and voltage-gated H^+ channel1 removes intracellular H^+. K^+ channel helps maintain the balance of the membrane potential and other channels mediate the transport of HCO_3^-. The pH and HCO_3^- rising activates CatSper promoting calcium ion, priming of the spermatozoon. The alkaline intracellular environment and Ca^{2+} trigger the atypical soluble adenylyl cyclase (SAC or SACY), which generates cAMP, that by its turn activates protein kinase A (PKA, Finkelstein et al. 2020). PKA leads the activation of sperm tyrosine kinases to trigger a cascade of protein phosphorylation involved in sperm motility. The tyrosine phosphorylation represents the hallmark of sperm capacitation ad it is influenced also by pathways involving receptor tyrosine kinase non-receptor protein tyrosine kinase (Pereira et al. 2017; Suarez et al. 2007).

Pharmacological treatments have demonstrated that capacitation is associated with ryanodine receptors, CatSper, and store-operated calcium channels (Vaquer et al. 2020).

3.6 Conclusions

The integrity of all the sperm structures is required for oocyte fertilization and embryo development. Therefore, the study of morphology and the molecular composition of normal human spermatozoa is the base for understanding the sperm alterations (Chap. 5) caused by genetic, environmental, and lifestyle factors or by different pathologies affecting the male reproductive tract (Chap. 4). In addition, human spermatozoa represent an in vitro cell model to test the effects of different compounds. The activity of these compounds is verified by analyzing the functional status of the different sperm structures, such as acrosome, chromatin, motility, vitality.

Spermatozoa as Models for In Vitro Studies

Spermatozoa represent a reliable model for in vitro studies for several reasons:

- fertile men are an important source of cellular material,
- sperm motility is an index of cell function and a marker of compound toxicity,
- spermatozoa are highly influenced by external environment for their low level of transcription and the lack of DNA-repairing activity,
- human spermatozoa have reduced capacity to counteract oxidative stress since the antioxidants are present in the seminal plasma.

Therefore, spermatozoa have been used in vitro as a monitor to test antioxidant, toxic compounds, and natural substances with potential contraceptive activity (Fig. 3.4).

When these in vitro protocols are applied, several endpoints should be assessed to understand how the different sperm organelles and structures react to the treatment (Fig. 3.4).

- The motility can be easily measured with an optical microscope or computerized systems providing more objective evaluations.
- The acrosomal shape can be studied with several molecular biomarkers such as acrosin, or fluorescent lectins that bind particular carbohydrates.
- The DNA integrity can be assessed with a single-cell gel electrophoresis test (comet assay) and terminal deoxynucleotidyl transferase deoxyuridine triphosphate (dUTP) nick-end labeling (TUNEL). In addition, indirect tests based on the use of acridine orange assess DNA vulnerability to denaturation in an acidic environment.
- Commercially available probes for membrane mitochondrial potential can be used for the study of mitochondria.
- Sperm vitality tests show if the immotile spermatozoa are dead or alive. This aspect is crucial in order to determine whether the studied compounds affect sperm motility or have the ability to kill the spermatozoa. The colorimetric tests frequently used to assess sperm vitality are based on the status of the plasma membrane: if the membrane is damaged, the spermatozoa are dead and stained, alive spermatozoa remain unstained.

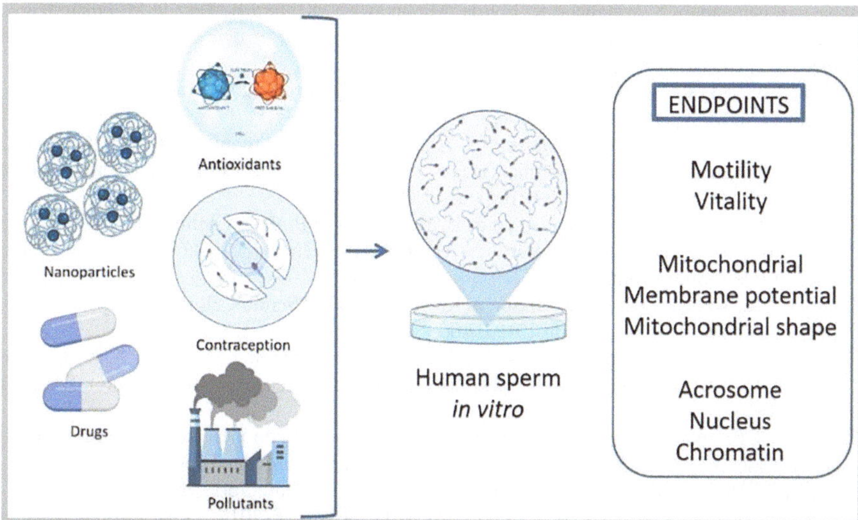

Fig. 3.4 Diagram showing the role of human spermatozoa as cell model to test in vitro the effects of different compounds. The evaluation of sperm motility, vitality, mitochondrial membrane potential and shape, acrosome and chromatin/DNA reveals the beneficial or toxic effects of the tested substances

References

Aarabi M, Balakier H, Bashar S, Moskovtsev SI, Sutovsky P, Librach CL, Oko R (2014) Sperm content of postacrosomal WW binding protein is related to fertilization outcomes in patients undergoing assisted reproductive technology. Fertil Steril 102(2):440–447. https://doi.org/10.1016/j.fertnstert.2014.05.003

Aitken RJ (2017) Reactive oxygen species as mediators of sperm capacitation and pathological damage. Mol Reprod Dev 84(10):1039–1052. https://doi.org/10.1002/mrd.22871

Aitken RJ, Gibb Z, Baker MA, Drevet J, Gharagozloo P (2016) Causes and consequences of oxidative stress in spermatozoa. Reprod Fertil Dev 28(1–2):1–10. https://doi.org/10.1071/RD15325

Alberts B (2015) Molecular biology of the cell. In: Garland Science, 6th edn. Taylor and Francis Group, New York

Amaral A, Ramalho-Santos J, St John JC (2007) The expression of polymerase gamma and mitochondrial transcription factor A and the regulation of mitochondrial DNA content in mature human sperm. Hum Reprod 22(6):1585–1596. https://doi.org/10.1093/humrep/dem030

Amaral A, Lourenço B, Marques M, Ramalho-Santos J (2013) Mitochondria functionality and sperm quality. Reproduction 146(5):R163–R174. https://doi.org/10.1530/REP-13-0178

Avidor-Reiss T, Fishman EL (2019) It takes two (centrioles) to tango. Reproduction 157(2):R33–R51

Avidor-Reiss T, Carr A, Fishman EL (2020) The sperm centrioles. Mol Cell Endocrinol 518:110987

Aydin H, Sultana A, Li S, Thavalingam A, Lee JE (2016) Molecular architecture of the human sperm IZUMO1 and egg JUNO fertilization complex. Nature 534(7608):562–565. https://doi.org/10.1038/nature18595

Baba T, Kashiwabara S, Watanabe K, Itoh H, Michikawa Y, Kimura K, Takada M, Fukamizu A, Arai Y (1989) Activation and maturation mechanisms of boar acrosin zymogen based on the deduced primary structure. J Biol Chem 264:11920–11927

Baccetti B (1984) Lo spermatozoo umano. In: Conti C, Isidori A (eds) Trattato di Andrologia. Piccin, Padova, pp 157–197

Beurois J, Cazin C, Kherraf ZE, Martinez G, Celse T, Touré A, Arnoult C, Ray PF, Coutton C (2020) Genetics of teratozoospermia: Back to the head. Best Pract Res Clin Endocrinol Metab 34(6):101473. https://doi.org/10.1016/j.beem.2020.101473

Boguenet M, Bouet PE, Spiers A, Reynier P, May-Panloup P (2021) Mitochondria: their role in spermatozoa and in male infertility. Hum Reprod Update 27(4):697–719. https://doi.org/10.1093/humupd/dmab001

Bohacek J, Rassoulzadegan M (2020) Sperm RNA: Quo vadis? Semin Cell Dev Biol 97:123–130. https://doi.org/10.1016/j.semcdb.2019.07.005

Brenker C, Goodwin N, Weyand I, Kashikar ND, Naruse M, Krahling M, Muller A, Kaupp UB, Strunker T (2012) The Catsper channel: a polymodal chemosensor in human sperm. EMBO J 31(7):1654–1665

Carracedo S, Briand-Amirat L, Dordas-Perpinyà M, Ramos Escuredo Y, Delcombel R, Sergeant N, Delehedde M (2022) ProAKAP4 protein marker: towards a functional approach to male fertility. Anim Reprod Sci 247:107074. https://doi.org/10.1016/j.anireprosci.2022.107074

Chai R-R, Chen G-W, Shi H-J, Wai-Sum O, PA M-DL, Chen H (2017) Prohibitin involvement in the generation of mitochondrial superoxide at complex I in human sperm. J Cell Mol Med 21:121–129

Chemes HE (2017) Sperm ultrastructure in fertile man and male sterility. Revisiting teratozoospermia. In: De Jonge CJ, Barratt CLR (eds) The sperm cell, 2nd edn. Cambridge University Press, Cambridge, pp 36–58

Collodel G, Castellini C, Lee JC, Signorini C (2020) Relevance of fatty acids to sperm maturation and quality. Oxidative Med Cell Longev 2020:7038124. https://doi.org/10.1155/2020/7038124

Corzett M, Mazrimas J, Balhorn R (2002) Protamine 1: protamine 2 stoichiometry in the sperm of eutherian mammals. Mol Reprod Dev 61:519–527. https://doi.org/10.1002/mrd.10105

Davis FM, Goulding EH, D'Agostin DM, Janardhan KS, Cummings CA, Bird GS, Eddy EM, Putney JW (2016) Male infertility in mice lacking the store-operated ca(2+) channel Orai1. Cell Calcium 59(4):189–197. https://doi.org/10.1016/j.ceca.2016.02.007

de la Iglesia A, Jodar M, Oliva R, Castillo J (2023) Insights into the sperm chromatin and implications for male infertility from a protein perspective. WIREs Mech Dis 15(2):e1588. https://doi.org/10.1002/wsbm.1588

Drevet JR, Hallak J, Nasr-Esfahani MH, Aitken RJ (2022) Reactive oxygen species and their consequences on the structure and function of mammalian spermatozoa. Antioxid Redox Signal 37(7–9):481–500. https://doi.org/10.1089/ars.2021.0235

Du Plessis S, Agarwal A, Mohanty G, van der Linde M (2015) Oxidative phosphorylation versus glycolysis: what fuel do spermatozoa use? Asian J Androl 17:230

Eddy EM (2007) The scaffold role of the fibrous sheath. Soc Reprod Fertil Suppl 65:45–62

Eddy EM, Toshimori K, O'Brien DA (2003) Fibrous sheath of mammalian spermatozoa. Microsc Res Tech 61(1):103–115. https://doi.org/10.1002/jemt.10320

Elkhatib RA, Paci M, Longepied G, Saias-Magnan J, Courbière B, Guichaoua MR, Lévy N, Metzler-Guillemain C, Mitchell MJ (2017) Homozygous deletion of SUN5 in three men with decapitated spermatozoa. Hum Mol Genet 26(16):3167–3171. https://doi.org/10.1093/hmg/ddx200

Esmaeili V, Shahverdi AH, Moghadasian MH, Alizadeh AR (2015) Dietary fatty acids affect semen quality: a review. Andrology 3(3):450–461. https://doi.org/10.1111/andr.12024

Fawcett DW (1970) A comparative view of sperm ultrastructure. Biol Reprod 2:90–127

Fawcett DW (1975) The mammalian spermatozoon. Dev Biol 44:394–436

Ferreira AF, Santiago J, Silva JV, Oliveira PF, Fardilha M (2022) PP1, PP2A and PP2B interplay in the regulation of sperm motility: lessons from protein phosphatase inhibitors. Int J Mol Sci 23(23):15235. https://doi.org/10.3390/ijms232315235

Finkelstein M, Etkovitz N, Breitbart H (2020) Ca2+ signaling in mammalian spermatozoa. Mol Cell Endocrinol 516:110953. https://doi.org/10.1016/j.mce.2020.110953

Fishman EL, Jo K, Nguyen QPH, Kong D, Royfman R, Cekic AR, Khanal S, Miller AL, Simerly C, Schatten G, Loncarek J, Mennella V, Avidor-Reiss T (2018) A novel atypical sperm centriole is functional during human fertilization. Nat Commun 9(1):2210. https://doi.org/10.1038/s41467-018-04678-8

Ford WC (2006) Glycolysis and sperm motility: does a spoonful of sugar help the flagellum go round? Hum Reprod Update 12(3):269–274. https://doi.org/10.1093/humupd/dmi053

Freitas MJ, Vijayaraghavan S, Fardilha M (2017) Signaling mechanisms in mammalian sperm motility. Biol Reprod 96(1):2–12. https://doi.org/10.1095/biolreprod.116.144337

Fujita A, Nakamura K, Kato T, Watanabe N, Ishizaki T, Kimura K, Mizoguchi A, Narumiya S (2000) Ropporin, a sperm-specific binding protein of rhophilin, that is localized in the fibrous sheath of sperm flagella. J Cell Sci 113(Pt 1):103–112. https://doi.org/10.1242/jcs.113.1.103

Gadêlha H, Hernández-Herrera P, Montoya F, Darszon A, Corkidi G (2020) Human sperm uses asymmetric and anisotropic flagellar controls to regulate swimming symmetry and cell steering. Sci Adv 6(31):eaba5168. https://doi.org/10.1126/sciadv.aba5168

Garolla A, Šabović I, Tescari S, De Toni L, Menegazzo M, Cosci I, De Filippis V, Giarola M, Foresta C (2018) Impaired sperm function in infertile men relies on the membrane sterol pattern. Andrology 6(2):325–334. https://doi.org/10.1111/andr.12468

Gat I, Orvieto R (2017) This is where it all started- the pivotal role of PLCζ within the sophisticated process of mammalian reproduction: a systemic review. Basic Clin Androl 27:9. https://doi.org/10.1186/s12610-017-0054-y

Gibb Z, Griffin RA, Aitken RJ, De Iuliis GN (2020) Functions and effects of reactive oxygen species in male fertility. Anim Reprod Sci 220:106456. https://doi.org/10.1016/j.anireprosci.2020.106456

Gómez-Torres MJ, Sáez-Espinosa P, Manzano-Santiago P, Robles-Gómez L, Huerta-Retamal N, Aizpurua J (2022) Sperm adhesion molecule 1 (SPAM1) distribution in selected human sperm by hyaluronic acid test. Biomedicines 10(10):2553. https://doi.org/10.3390/biomedicines10102553

Gómez-Torres MJ, Luna-Romero J, Fernández-Colom PJ, Aizpurua J, Avilés M, Romero A (2023) Human sperm head vacuoles are related to nuclear-envelope invaginations. Int J Mol Sci 24(12):10027. https://doi.org/10.3390/ijms241210027

Gu N-H, Zhao W-L, Wang G-S, Sun F (2019) Comparative analysis of mammalian sperm ultrastructure reveals relationships between sperm morphology, mitochondrial functions and motility. Reprod Biol Endocrinol 17:66

Hansen JN, Rassmann S, Jikeli JF, Wachten D (2018) SpermQ⁻A simple analysis software to comprehensively study flagellar beating and sperm steering. Cells 8(1):10. https://doi.org/10.3390/cells8010010

Hao J, Chen M, Ji S, Wang X, Wang Y, Huang X, Yang L, Wang Y, Cui X, Lv L, Liu Y, Gao F (2014) Equatorin is not essential for acrosome biogenesis but is required for the acrosome reaction. Biochem Biophys Res Commun 444(4):537–542. https://doi.org/10.1016/j.bbrc.2014.01.080

Hetherington L, Schneider EK, Scott C, DeKretser D, Muller CH, Hondermarck H, Velkov T, Baker MA (2016) Deficiency in outer dense Fiber 1 is a marker and potential driver of idiopathic male infertility. Mol Cell Proteomics 15(12):3685–3693. https://doi.org/10.1074/mcp.M116.060343

Hirose M, Honda A, Fulka H, Tamura-Nakano M, Matoba S, Tomishima T, Mochida K, Hasegawa A, Nagashima K, Inoue K, Ohtsuka M, Baba T, Yanagimachi R, Ogura A (2020) Acrosin is essential for sperm penetration through the zona pellucida in hamsters. Proc Natl Acad Sci USA 117(5):2513–2518. https://doi.org/10.1073/pnas.1917595117

Ho HC (2010) Redistribution of nuclear pores during formation of the redundant nuclear envelope in mouse spermatids. J Anat 216(4):525–532. https://doi.org/10.1111/j.1469-7580.2009.01204.x

Hoghoughi N, Barral S, Vargas A, Rousseaux S, Khochbin S (2018) Histone variants: essential actors in male genome programming. J Biochem 163(2):97–103. https://doi.org/10.1093/jb/mvx079

Inaba K (2011) Sperm flagella: comparative and phylogenetic perspectives of protein components. Mol Hum Reprod 17(8):524–538. https://doi.org/10.1093/molehr/gar034

Inoue N, Ikawa M, Isotani A, Okabe M (2005) The immunoglobulin superfamily protein Izumo is required for sperm to fuse with eggs. Nature 434(7030):234–238. https://doi.org/10.1038/nature03362

Ito C, Toshimori K (2016) Acrosome markers of human sperm. Anat Sci Int 91(2):128–142. https://doi.org/10.1007/s12565-015-0323-9

Kadam K, Khole VV, Ghosalkar K, Jagtap D, Yarramala DS, Ramachandran B (2020) Thiol based mechanism internalises interacting partners to outer dense fibers in sperm. Free Radic Biol Med 148:170–181. https://doi.org/10.1016/j.freeradbiomed.2019.12.031

Kashir J, Heindryckx B, Jones C, De Sutter P, Parrington J, Coward K (2010) Oocyte activation, phospholipase C zeta and human infertility. Hum Reprod Update 16(6):690–703. https://doi.org/10.1093/humupd/dmq018

Kashir J, Nomikos M, Swann K, Lai FA (2015) PLCζ or PAWP: revisiting the putative mammalian sperm factor that triggers egg activation and embryogenesis. Mol Hum Reprod 21(5):383–388. https://doi.org/10.1093/molehr/gav009

Kumar N, Singh AK (2021) The anatomy, movement, and functions of human sperm tail: an evolving mystery. Biol Reprod 104(3):508–520

Lass A, Belluzzi A (2019) Omega 3 polyunsaturated fatty acids and IVF treatment. Reprod Biomed Online 38(1):95–99. https://doi.org/10.1016/j.rbmo.2018.10.008

Lehti MS, Sironen A (2017) Formation and function of sperm tail structures in association with sperm motility defects. Biol Reprod 97:522–536

Lenzi A, Picardo M, Gandini L, Dondero F (1996) Lipids of the sperm plasma membrane: from polyunsaturated fatty acids considered as markers of sperm function to possible scavenger therapy. Hum Reprod Update 2(3):246–256. https://doi.org/10.1093/humupd/2.3.246

Li YF, He W, Mandal A, Kim YH, Digilio L, Klotz K, Flickinger CJ, Herr JC, Herr JC (2011) CABYR binds to AKAP3 and Ropporin in the human sperm fibrous sheath. Asian J Androl 13(2):266–274. https://doi.org/10.1038/aja.2010.149

Linck RW, Chemes H, Albertini DF (2016) The axoneme: the propulsive engine of spermatozoa and cilia and associated ciliopathies leading to infertility. J Assist Reprod Genet 33(2):141–156. https://doi.org/10.1007/s10815-016-0652-1

Liu S, Sharma U (2023) Sperm RNA payload: implications for intergenerational epigenetic inheritance. Int J Mol Sci 24(6):5889. https://doi.org/10.3390/ijms24065889

Manandhar G, Simerly C, Schatten G (2000) Highly degenerated distal centrioles in rhesus and human spermatozoa. Hum Reprod 15:256–263. https://doi.org/10.1093/humrep/15.2.256

Martin-Hidalgo D, Hurtado de Llera A, Calle-Guisado V, Gonzalez-Fernandez L, Garcia-Marin L, Bragado M (2018) AMPK function in mammalian spermatozoa. Int J Mol Sci 19:3293

McSwiggin HM, O'Doherty AM (2018) Epigenetic reprogramming during spermatogenesis and male factor infertility. Reproduction 156(2):R9–R21. https://doi.org/10.1530/REP-18-0009

Miki K, Willis WD, Brown PR, Goulding EH, Fulcher KD, Eddy EM (2002) Targeted disruption of the Akap4 gene causes defects in sperm flagellum and motility. Dev Biol 248(2):331–342. https://doi.org/10.1006/dbio.2002.0728

Miki K, Qu W, Goulding EH, Willis WD, Bunch DO, Strader LF, Perreault SD, Eddy EM, O'Brien DA (2004) Glyceraldehyde 3-phosphate dehydrogenase-S, a sperm-specific glycolytic enzyme, is required for sperm motility and male fertility. Proc Natl Acad Sci USA 101:16501–16506

Mohri H (1968) Amino-acid composition of "tubulin" constituting microtubules of sperm flagella. Nature 217:1053–1054

Moretti E, Noto D, Corsaro R, Collodel G (2023a) Focus on centrin in normal and altered human spermatozoa. Syst Biol Reprod Med 69(3):175–187. https://doi.org/10.1080/19396368.2023.2181115

Moretti E, Signorini C, Corsaro R, Giamalidi M, Collodel G (2023b) Human sperm as an in vitro model to assess the efficacy of antioxidant supplements during sperm handling: A narrative review. Antioxidants (Basel) 12(5):1098. https://doi.org/10.3390/antiox12051098

Moritz L, Hammoud SS (2022) The art of packaging the sperm genome: molecular and structural basis of the histone-to-protamine exchange. Front Endocrinol (Lausanne) 13:895502. https://doi.org/10.3389/fendo.2022.895502

Mortimer D (2018) The functional anatomy of the human spermatozoon: relating ultrastructure and function. Mol Hum Reprod 24(12):567–592. https://doi.org/10.1093/molehr/gay040

Nomikos M, Sanders JR, Kashir J, Sanusi R, Buntwal L, Love D, Ashley P, Sanders D, Knaggs P, Bunkheila A, Swann K, Lai FA (2015) Functional disparity between human PAWP and PLCζ in the generation of Ca2+ oscillations for oocyte activation. Mol Hum Reprod 21(9):702–710. https://doi.org/10.1093/molehr/gav034

O'Flaherty C (2019) Orchestrating the antioxidant defenses in the epididymis. Andrology 7(5):662–668. https://doi.org/10.1111/andr.12630

Oliva R (2006) Protamines and male infertility. Hum Reprod Update 12:417–435. https://doi.org/10.1093/humupd/dml009

Paul N, Talluri TR, Nag P, Kumaresan A (2021) Epididymosomes: A potential male fertility influencer. Andrologia 53(9):e14155. https://doi.org/10.1111/and.14155

Pelech S, Vieira S, Rebelo S, Korrodi-Gregorio L, Sousa M, Barros A, Silva V, da Cruz e Silva OA, da Cruz e Silva EF (2013) Omics of human sperm: profiling protein phosphatases. OMICS 17(9):460–472. https://doi.org/10.1089/omi.2012.0119

Pelloni M, Paoli D, Majoli M, Pallotti F, Carlini T, Lenzi A, Lombardo F (2018) Molecular study of human sperm RNA: Ropporin and CABYR in asthenozoospermia. J Endocrinol Investig 41(7):781–787. https://doi.org/10.1007/s40618-017-0804-x

Pereira R, Sá R, Barros A, Sousa M (2017) Major regulatory mechanisms involved in sperm motility. Asian J Androl 19(1):5–14. https://doi.org/10.4103/1008-682X.167716

Rahban R, Nef S (2020) CatSper: the complex main gate of calcium entry in mammalian spermatozoa. Mol Cell Endocrinol 518:110951. https://doi.org/10.1016/j.mce.2020.110951

Ribas-Maynou J, Nguyen H, Wu H, Ward WS (2022) Functional aspects of sperm chromatin organization. Results Probl Cell Differ 70:295–311. https://doi.org/10.1007/978-3-031-06573-6_10

Saleh A, Kashir J, Thanassoulas A, Safieh-Garabedian B, Lai FA, Nomikos M (2020) Essential role of sperm-specific PLC-zeta in egg activation and male factor infertility: an update. Front Cell Dev Biol 8:28. https://doi.org/10.3389/fcell.2020.00028

Santiago J, Silva JV, Howl J, Santos MAS, Fardilha M (2021) All you need to know about sperm RNAs. Hum Reprod Update 28:67–91

Schröter S, Osterhoff C, McArdle W, Ivell R (1999) The glycocalyx of the sperm surface. Hum Reprod Update 5(4):302–313. https://doi.org/10.1093/humupd/5.4.302

Shen YR, Wang HY, Tsai YC, Kuo YC, Wu SR, Wang CY, Kuo PL (2020) The SEPT12 complex is required for the establishment of a functional sperm head-tail junction. Mol Hum Reprod 26(6):402–412

Singh AP, Rajender S (2015) CatSper channel, sperm function and male fertility. Reprod Biomed Online 30(1):28–38. https://doi.org/10.1016/j.rbmo.2014.09.014

Sosnik J, Miranda PV, Spiridonov NA et al (2009) Tssk6 is required for Izumo relocalization and gamete fusion in the mouse. J Cell Sci 122:2741–2749

Suarez SS, Marquez B, Harris TP, Schimenti JC (2007) Different regulatory systems operate in the midpiece and principal piece of the mammalian sperm flagellum. Soc Reprod Fertil Suppl 65:331–334

Sun XH, Zhu YY, Wang L, Liu HL, Ling Y, Li ZL, Sun LB (2017) The Catsper channel and its roles in male fertility: a systematic review. Reprod Biol Endocrinol 15(1):65. https://doi.org/10.1186/s12958-017-0281-2

Tachiwana H, Osakabe A, Kimura H, Kurumizaka H (2008) Nucleosome formation with the testis-specific histone H3 variant, H3t, by human nucleosome assembly proteins in vitro. Nucleic Acids Res 36:2208–2218. https://doi.org/10.1093/nar/gkn060

Tarozzi N, Nadalini M, Coticchio G, Zacà C, Lagalla C, Borini A (2021) The paternal toolbox for embryo development and health. Mol Hum Reprod 27(7):gaab042

Toshimori K, Saxena DK, Tanii I, Yoshinaga K (1998) An MN9 antigenic molecule, equatorin, is required for successful sperm–oocyte fusion in mice. Biol Reprod 59:22–29

Vaquer CC, Suhaiman L, Pavarotti MA, De Blas GA, Belmonte SA (2020) Ceramide induces a multicomponent intracellular calcium increase triggering the acrosome secretion in human sperm. Biochim Biophys Acta, Mol Cell Res 1867:118704. https://doi.org/10.1016/j.bbamcr.2020.118704

Wang M-J, Ou J-X, Chen G-W, Wu J-P, Shi H-J, Wai-Sum O, PA M-DL, Chen H (2012) Does prohibitin expression regulate sperm mitochondrial membrane potential, sperm motility, and male fertility? Antioxid Redox Signal 17:513–519

Wang X, Qian H, Huang X, Li J, Zhang J, Zhu N, Chen H, Zhu C, Wang J, Zhang P, Zhang P, Jin C, Het G (2018) UCP2 mitigates the loss of human spermatozoa motility by promoting mROS elimination. Cell Physiol Biochem 50:952–962

Ward WS (2010) Function of sperm chromatin structural elements in fertilization and development. Mol Hum Reprod 16:30–36. https://doi.org/10.1093/molehr/gap080

Wei YL, Yang WX (2018) The acroframosome-acroplaxome-manchette axis may function in sperm head shaping and male fertility. Gene 660:28–40. https://doi.org/10.1016/j.gene.2018.03.059

Wolkowicz MJ, Digilio L, Klotz K, Shetty J, Flickinger CJ, Herr JC (2008) Equatorial segment protein (ESP) is a human alloantigen involved in sperm-egg binding and fusion. J Androl 29(3):272–282. https://doi.org/10.2164/jandrol.106.000604

Yan C, Duanmu X, Zeng L, Liu B, Song Z (2019) Mitochondrial DNA: distribution, mutations, and elimination. Cells 8(4):379. https://doi.org/10.3390/cells8040379

Yu W, Li Y, Chen H, Cui Y, Situ C, Yao L, Zhang X, Lu S, Liu L, Li L, Ren J, Guo Y, Huo Z, Chen Y, Li H, Jiang T, Gu Y, Wang C, Zhu T, Li Y, Hu Z, Guo X (2023) STK33 phosphorylates fibrous sheath protein AKAP3/4 to regulate sperm flagella assembly in spermiogenesis. Mol Cell Proteomics 22(6):100564. https://doi.org/10.1016/j.mcpro.2023.100564

Yuan S, Stratton CJ, Bao J, Zheng H, Bhetwal BP, Yanagimachi R, Yan W (2015) Spata6 is required for normal assembly of the sperm connecting piece and tight head-tail conjunction. Proc Natl Acad Sci USA 112(5):E430–E439. https://doi.org/10.1073/pnas.1424648112

Zabeo D, Heumann JM, Schwartz CL, Suzuki-Shinjo A, Morgan G, Widlund PO, Höög JL (2018) A lumenal interrupted helix in human sperm tail microtubules. Sci Rep 8(1):2727. https://doi.org/10.1038/s41598-018-21165-8

Zalata AA, Christophe AB, Depuydt CE, Schoonjans F, Comhaire FH (1998) The fatty acid composition of phospholipids of spermatozoa from infertile patients. Mol Hum Reprod 4(2):111–118. https://doi.org/10.1093/molehr/4.2.111

Zaneveld LJD, De Jonge CJ (1991) Mammalian sperm acrosomal enzymes and the acrosome reaction. In: Dunbar BS, O'Rand MG (eds) A comparative overview of mammalian fertilization. Springer, Boston, MA. https://doi.org/10.1007/978-1-4757-8982-9_4

Zerbinati C, Caponecchia L, Rago R, Leoncini E, Bottaccioli AG, Ciacciarelli M, Pacelli A, Salacone P, Sebastianelli A, Pastore A, Palleschi G, Boccia S, Carbone A, Iuliano L (2016) Fatty acids profiling reveals potential candidate markers of semen quality. Andrology 4(6):1094–1101. https://doi.org/10.1111/andr.12236

Zhao M, Shirley CR, Hayashi S, Marcon L, Mohapatra B, Suganuma R, Behringer RR, Boissonneault G, Yanagimachi R, Meistrich ML (2004) Transition nuclear proteins are required for Normal chromatin condensation and functional sperm development. Genesis 38:200–213. https://doi.org/10.1002/gene.20019

Zhao W, Li Z, Ping P, Wang G, Yuan X, Sun F (2018) Outer dense fibers stabilize the axoneme to maintain sperm motility. J Cell Mol Med 22(3):1755–1768. https://doi.org/10.1111/jcmm.13457

Zhu F, Liu C, Wang F, Yang X, Zhang J, Wu H, Zhang Z, He X, Zhang Z, Zhou P, Wei Z, Shang Y, Wang L, Zhang R, Ouyang YC, Sun QY, Cao Y, Li W (2018) Mutations in PMFBP1 cause Acephalic spermatozoa syndrome. Am J Hum Genet 103(2):188–199. https://doi.org/10.1016/j.ajhg.2018.06.010

Factors Affecting Human Spermatogenesis

4

Contents

Core Messages

The male factor appears to be responsible for 30%–50% of infertility. There are many causes of male subfertility, which can be traced back to congenital, acquired, or idiopathic factors.

This chapter deals first with the relationship between oxidative stress and spermatozoa, then with some of the most frequent factors affecting male fertility. The factors include varicocele, urogenital infections, inflammation, obesity, genetic factors, pollutants, and lifestyle. Recently, it has been suggested that the spermatozoon accumulates epigenetic errors because of the constant mitotic and meiotic cycles during the spermatogenetic process. A better knowledge of the molecular and genetic factors involved in pathological reproductive conditions allows a deep investigation into the etiology of male factor infertility and a possible improvement of patients' chances of reproductive success.

E. Moretti, G. Collodel, *The Human Spermatozoon*,
https://doi.org/10.1007/978-3-031-97868-5_4

Abbreviations

AR	Androgen receptor
ARTs	Assisted reproductive techniques
AZF	Azoospermia factor
CAVD	Congenital absence of the vas deferens
DAZ	Deleted in azoospermia
DBY	DEAD-Box Helicase 3 Y-Linked
F_2-IsoPs	F_2-Isoprostanes
FSH	Follicle stimulating hormone
IL	Interleukin
LH	Luteinizing hormone
LPO	Lipid peroxidation
NOA	Non-obstructive azoospermia
OS	Oxidative stress
PRY	PTP-BL related on the Y chromosome
PUFAs	Polyunsaturated fatty acids
RBMY	RNA-binding motif Y chromosome
ROS	Reactive oxygen species
RPL	Recurrent pregnancy loss
SPMs	Specialized pro-resolving mediators

4.1 Several Factors Affecting Human Spermatogenesis

The decline of fertility in modern society is a significant worldwide concern, and the reasons behind this phenomenon are complex and difficult to understand. Infertility is a well-recognized condition defined as the inability of a couple to conceive after 12 months of regular, unprotected sexual intercourses (WHO 2021). It is one of the major health issues affecting about 40%–50% of couples around the world, with the male factor accounting for about 50% of the cases. More precisely, as reported by the World Health Organization (WHO 2021), approximately 1 in 6 individuals worldwide experience infertility (Baskaran et al. 2021; Kumar and Singh 2015; Njagi et al. 2023). Male infertility seems to be more common in Central/Eastern Europe, the USA, and Australia (Agarwal et al. 2015), and the prevalence of infertility is higher among those delaying parenthood (Datta et al. 2016).

Male infertility is a multifactorial pathological condition (Fig. 4.1). The most relevant factors involved in male infertility are (Fainberg and Kashanian 2019; Krausz and Riera-Escamilla 2018):

- Anatomical alterations: varicocele, cryptorchidism, anorchia (congenital absence of testes), congenital absence of vas deferens (CAVD), obstructions of the ducts of the male genital tract,
- bacterial and viral infections and inflammations,
- genetic conditions such as Klinefelter syndrome, systematic sperm defects, karyotype alterations, and microdeletions of Y chromosome,

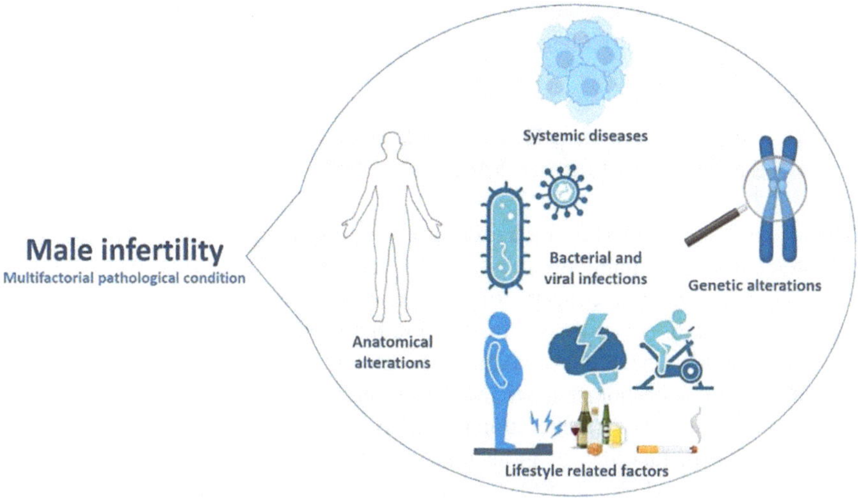

Fig. 4.1 Schematic diagram of several conditions affecting human male fertility

- cancer, systemic diseases as diabetes,
- lifestyle-related factors such as smoking, drinking, drug abuse, high energy-based diet, obesity, excessive physical activity, and psychological stress negatively influence the male reproductive potential (Eisenberg et al. 2023).

In addition, the use of medical therapies, the exposure to endocrine disruptors and dysfunctions of hypothalamic-pituitary axis can be responsible for male infertility. However, in about 30% of cases, male infertility (Fainberg and Kashanian 2019) remains undiagnosed and is categorized as *sine causa* (idiopathic infertility), thereby significantly limiting the available strategies for treatment.

It is important to note that the general health condition of a man is closely related with his fertility status. For example, men with abnormal semen parameters have a 20-fold greater incidence to develop malignancy. Ventimiglia et al. (2015) analyzed a cohort of 2100 infertile patients and observed that a compromised general health status is associated with impaired male reproductive function.

One of the most important acquisitions in the field of male infertility is the relevant role of oxidative stress (OS), which can be a causative factor in 30%–80% of infertile men (Agarwal et al. 2019).

Indeed, OS has emerged as a common finding in many reproductive pathologies as varicocele, genitourinary infections, inflammation, and in case of exposure to environmental toxicants, heavy metals, and in the presence of uncorrected lifestyle factors such as smoking (Aitken et al. 2014).

OS-induced male infertility represents an issue of great concern for clinicians and scientists because, from a clinical perspective, OS is a contributing agent of reduced fertilization rates, defective embryonic development, recurrent miscarriages, and overall success rate of assisted reproductive techniques (ARTs).

This chapter deals first with the relationship between OS and spermatozoa, then with some of the most frequent factors affecting male fertility.

4.1.1 Interplay Among Oxidative Stress, Reactive Oxygen Species and Human Spermatozoa

The idea of "homeostasis," which is described as a self-regulating process by which a biological system maintains stability while adapting to changing external conditions, is one of the core concepts of biological processes. Therefore, an imbalance in the major factors or processes involved in homeostatic regulation might have an impact on health. One of the key regulatory mechanisms involved in physiology and pathology is oxygen homeostasis (Moretti et al. 2023). The detrimental effects of oxygen are mainly due to generation of reactive oxygen species (ROS) that are partially reduced metabolites of molecular oxygen produced by different metabolic pathways (Sies et al. 2022). ROS act either as toxic metabolites or as regulator of essential cell signaling cascades. The balance between the rates of ROS production and the rates of their clearance by antioxidant defense mechanisms at physiological levels impact on the amount of ROS present in a system (Aitken et al. 2022; O'Flaherty 2020). Depending on the concentration of ROS, on the buffering capacity of the antioxidant system and on the interaction of ROS with cell targets, a condition of oxidative distress, named OS, may arise as a result of redox imbalance (Sies 2015). ROS include a wide category of molecules, such as oxygen free radicals such as superoxide anion (O_2^-), peroxide (O_2^{2-}) and hydroxyl radical (OH^-), non-radical species, and reactive nitrogen species (Masenga et al. 2023). Therefore, oxidative damage can occur due to non-specific reactions between oxidative agents and biological macromolecules as proteins, lipids, nucleic acids, and carbohydrates (Drevet et al. 2022). Spermatozoa are a perfect example of the oxygen paradox: they produce ROS because they need for several physiological processes, but when ROS generation overcomes the antioxidant capacity, damages to sperm macromolecules occur (Fig. 4.2).

ROS play a relevant role in sperm physiology, regulating several intracellular pathways and modulating the activation of different transcription factors (Mannucci et al. 2022).

Processes such as the compaction and folding of sperm chromatin, capacitation—during which spermatozoa undergo significant membrane changes and acquire hyperactivated motility—along with the acrosome reaction and subsequent fertilization, are all regulated by ROS (De Jonge 2017).

From a pathological point of view (Fig. 4.2), spermatozoa are the main target of ROS attack for several reasons: their membranes are rich in polyunsaturated fatty acids (PUFAs), they essentially lack the ability to repair DNA damages, and they are physiologically devoid of cytoplasm.

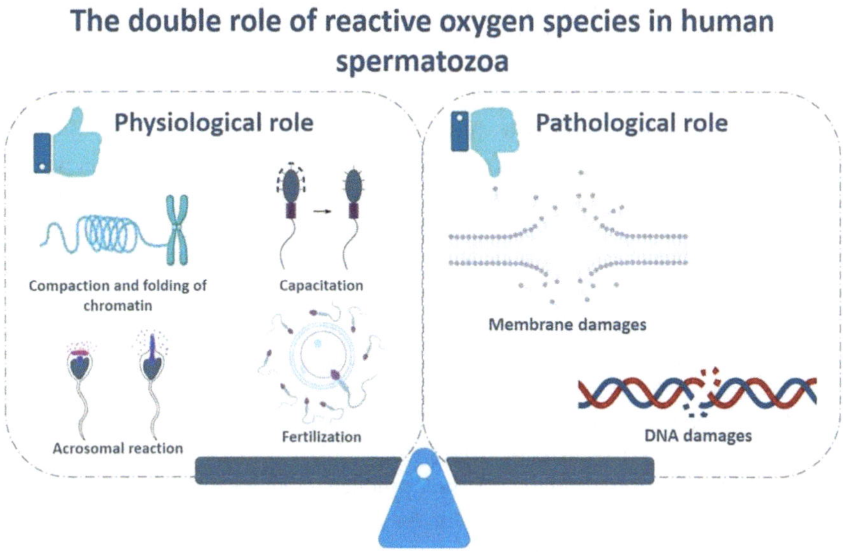

Fig. 4.2 Schematic diagram illustrating the dual role of reactive oxygen species (ROS). ROS are essential signaling molecules needed for sperm functions and damaging agents if their concentrations are not counterbalanced by an efficient antioxidant system

In addition, spermatozoa are equipped with a low amount of antioxidant enzymes such as glutathione peroxidase, catalase, superoxide dismutase, peroxiredoxins (O'Flaherty 2018) that have the ability to protect cells from oxidative damage. The antioxidant battery present in the seminal plasma balances the scarcity of spermatozoa antioxidant defenses (Lazzarino et al. 2019; Micheli et al. 2016). Seminal plasma can also contain cells other than mature spermatozoa; in particular, immature spermatozoa and leukocytes are the main contributors for ROS that are balanced by antioxidants of seminal plasma in physiological condition. However, when the amount of these cells exponentially grows, the antioxidant power of seminal plasma is not sufficient to counteract ROS, leading to OS (Aitken and Baker 2013). The link between alteration of redox balance and male infertility is well demonstrated (Aitken et al. 2022; Agarwal et al. 2018; Villaverde et al. 2019).

Due to the high content of PUFAs in their plasma membrane, spermatozoa are a preferential target of ROS insults that causes lipid peroxidation (LPO, Fig. 4.2). Upon initiation, LPO leads to dramatic remodeling of the composition and biophysical properties of sperm membranes, and, in the case of the mitochondria, this phenomenon causes a dissipation of membrane potential, electron leakage, increased ROS production, reduced capacity for energy production, and consequently loss of motility (Nowicka-Bauer and Nixon 2020).

As stated before, spermatozoa are quite silent from the transcriptional and translational point of view, and even though the chromatin is folded and compacted more than somatic cells, the DNA is a target of oxidative damage (Fig. 4.2) with potential harmful effects on reproductive outcome (Drevet and Aitken 2019). OS can act on sperm DNA in two different stages: during spermiogenesis, leading to defective sperm protamination and chromatin compaction and by a direct oxidative insult due to a ROS overproduction that can cause, among others, DNA fragmentation, base-pair modification, chromatin-cross linking.

Numerous studies have confirmed the harmful impact of OS on semen parameters such as sperm viability, motility, and fertilization potential as evidenced by the damages occurred to spermatozoa of infertile patients compared to fertile controls (Collodel et al. 2015; Micheli et al. 2016; Mannucci et al. 2022; Moretti et al. 2022; Ribeiro et al. 2022).

Much research agrees that OS is enhanced in altered conditions of the male reproductive system, leading to infertility such as varicocele (Wood et al. 2021), bacterial and viral infections, inflammation and leucocytospermia, and chronic pathologies such as diabetes and cancer (Agarwal et al. 2018; Jing et al. 2023; Leisegang et al. 2021; Wang et al. 2021).

ROS damage causes LPO and the first products generated consist of malondialdehyde, 4-hydroxynonenal, and acrolein. As secondary by-products of LPO, a range of prostaglandin-like molecules known as isoprostanoids have been detected in various mammalian tissues and fluids, indicating them as a gold standard biomarker of endogenous LPO (Collodel et al. 2022). Indeed, F_2-Isoprostanes (F_2-IsoPs) are common in human plasma and urine and recently have been used also as marker of OS in semen, thanks to their chemical stability and their measurability by non-invasive techniques (Milne 2017; Signorini et al. 2020).

Increased levels of F_2-IsoPs have been quantified in the semen of men with varicocele and urogenital infections, conditions which are commonly associated with high ROS levels and inflammation as previously analyzed (Longini et al. 2020; Moretti et al. 2022).

Acute inflammation is a rapid response that begins right after tissue damage. It typically resolves itself within a few days and is controlled by chemical mediators, such as breakdown enzymatic products of arachidonic acid (Dennis and Norris 2015). The persistence of chronic inflammation typically depends on the continuous presence of an underlying responsible factor. However, chronic inflammation may also arise from a failure in the resolution of inflammation, where pro-resolving signals are overwhelmed by proinflammatory stimuli. Therefore, the ineffectiveness in resolution inflammatory processes highlights the importance of evaluating inflammation resolution pathways, run by specialized pro-resolving mediators (SPMs, Collodel et al. 2022). SPMs dysregulation causes various pathological conditions, including male infertility, where inflammation negatively affects spermatogenesis. Specifically, Resolvin D1 (RvD1), derived from the docosahexaenoic acid, is associated with male infertility based on inflammation and reduced sperm quality, making it a potential biomarker for diagnosis and treatment of inflammatory male infertility.

Investigating the modulation of inflammatory pathways and evaluating SPMs could offer insights in the management of infertility conditions and in vitro fertilization outcomes.

4.2 Varicocele

Varicocele, generally diagnosed during a physical examination of the scrotum, is a dilation of the internal spermatic veins and pampiniform venous plexus draining the testis, caused by incompetent venous valves, leading to reflux and stasis of venous blood.

Varicocele is present in about 15% of adult males, but the prevalence increases to about 45% in the infertile male population (Clavijo et al. 2017; Damsgaard et al. 2016). According to the WHO, one out of four men with reduced semen quality is affected by varicocele (Bhattacharya et al. 2024).

Varicocele has emerged as an important cause of male infertility; however, it is a debated topic because men with varicocele are able to father children even without intervention (Jensen et al. 2017). The presence of this pathology plays a negative impact on the integrity of the blood-testis barrier down regulating the expression of E-cadherin and α-catenin, proteins involved in the junctional complexes of the Sertoli cells (Koksal et al. 2007) and interferes with the physiological removal of the residual sperm cytoplasm, resulting in defective sperm function (Zini et al. 2000).

Analyzing testicular biopsies of infertile men with varicocele, the most common findings observed account for low spermatozoa production, early and late maturation arrest, decreased diameter of seminiferous tubule, presence of only Sertoli cells with focal spermatogenesis, Leydig cell hyperplasia (Ustuner et al. 2015).

The exact mechanisms of actions by which the varicocele reduces male fertility remains somehow elusive (Neto et al. 2016); however, infertile men with varicocele benefit of surgical treatments from the point of view of semen parameters, pregnancy rates, and testosterone levels.

Different hypotheses have been suggested to explain the possible negative effect of varicocele on spermatogenesis, but no single theory explains how varicocele can affect spermatogenesis, and the most plausible mechanisms are extrapolated by animal models (Razi and Malekinejad 2015).

Among these mechanisms, testicular hypoperfusion and blood stasis in the microcirculatory vessels lead to cellular alterations characteristic of ischemia (Gat et al. 2005; Lee et al. 2006). Additionally, scrotal hyperthermia results in elevated temperatures that are detrimental to the spermatogenic process, inducing germ cell apoptosis and increased expression of heat shock proteins (Neto et al. 2016).

Recently, OS and inflammation (Fig. 4.3) were suggested as potential causes of detrimental effect of varicocele in male fertility since elevated ROS concentrations have been observed in semen of patients affected by this pathology (Ritchie and Ko 2021). In addition, high ROS level can compromise sperm quality, in particular sperm motility, sperm maturation (Collodel et al. 2018), and DNA integrity leading to a decreased fertility potential. In general, a concurrent reduction of OS markers and DNA fragmentation after varicocele repair is reported in the literature (Cannarella et al. 2024; Esteves et al. 2021; Kirby et al. 2016). In fact, while direct evidence on improving pregnancy rate is not available now, varicocelectomy has demonstrated promising results in relieving OS (Lai et al. 2022).

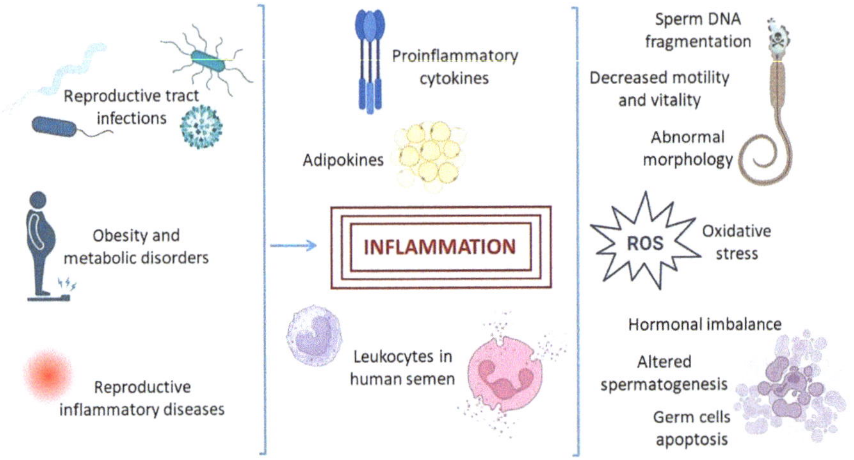

Fig. 4.3 Schematic diagram illustrating some common causes of inflammation of the male reproductive system such as infections, obesity and, in general pathologies with inflammatory background (left part of the figure). The central part shows some of the most important actors in inflammations. The last part of the figure indicates the damages induced by reactive oxygen species (ROS) overproduction as a consequence of inflammation

In a recent meta-analysis, Tian et al. (2023) reported the negative effect of varicocele on hormonal status; in fact, after surgical treatment the testis function was improved concomitantly with an increase of serum testosterone, and inhibin B levels and a decrease of serum follicle stimulating hormone (FSH) and luteinizing hormone (LH) concentration.

Varicocele is a pathology often observed in adolescent patients, and this is particularly complex issue for the future fertility perspectives. For example, the semen of adolescents with varicocele and low sperm parameters shows increased expression of semenogelin 1 and 2. Since semenogelins play a role in inhibiting motility and early capacitation, these results can explain reduced sperm motility observed in presence of varicocele (Neto et al. 2016; Zylbersztejn et al. 2013).

To determine which teenagers need therapy and when it should start, additional research employing sound methodology is required to prevent any negative impacts on fertility in the future (Mason and Clavijo 2023).

4.3 Infections and Inflammations

Sperm bacterial contamination is a frequent finding during human semen analysis; the presence of bacterial colony in the semen can contribute to the deterioration of the sperm quality of infertile men (Moretti et al. 2009a). A variety of microorganisms, in particular, viruses and bacteria have been identified as pathogens of the genitourinary system and can cause subfertility. Indeed, about 15% of the male infertility cases are related to bacterial genitourinary infections (Oghbaei et al.

2020). The presence of bacteria in concentrations greater than 1×10^3 bacteria/mL in the ejaculate is clinically regarded as a sign of an active infection and is called bacteriospermia (Farsimadan and Motamedifar 2020).

Most research has focused on the different parts of male genital tract that can be targets of infection such as testis, epididymis, and accessory glands; however, the studies on the harmful effects of the different microbial species on spermatozoa and their mechanism of action are still debated and under-studied (Oghbaei et al. 2020).

Infections of the male genital tract are mainly caused by sexually transmitted pathogens and here are reported the most common investigated bacterial microorganisms affecting semen quality and interfering with male fertility (Farsimadan and Motamedifar 2020):

(a) a group of atypical bacteria including, *Chlamydia trachomatis*, *Ureaplasma urealyticum*, *Ureaplasma parvum*, *Mycoplasma genitalium*, and *Mycoplasma hominis*,
(b) a group of aerobic bacteria including, *Escherichia coli*, *Staphylococcus aureus*, *Neisseria gonorrhoeae*, *Pseudomonas aeruginosa*, *Enterococcus faecalis*, and *Streptococcus agalactiae*,
(c) a group of anaerobic bacteria including, *Gardnerella vaginalis* and *Anaerococcus*.

Among these bacteria, *E. coli* and *C. trachomatis* are the most clinically relevant pathogens and thus are frequently used in animal studies to mimic the human condition (Hasan et al. 2022).

The research has demonstrated that human intestinal floras such *E. faecalis* and several intestinal and extra-intestinal pathogenic *E. coli* strains play an important role in the genitourinary infections (Oghbaei et al. 2020).

Several viral infections such as *Human Papilloma* virus (Garolla et al. 2024), *Hepatitis B* virus (Cheng et al. 2000) exert a detrimental role on spermatogenesis and spermatozoa affecting male infertility. In addition, *Human Immunodeficiency* virus-1, *Zika* virus, *Ebola* and *Marburg* viruses and others can infect the male reproductive tract of human and nonhuman primates via bloodstream. These viruses take a long time to spread throughout the organ. According to recent research, certain viruses may be stored in the testicular macrophages, which are essential for the initial infection and subsequent spread to other testicular cells (Hasan et al. 2022).

Even chronic systemic infections as those caused by *Helicobacter pylori*, *Mycoplasma pneumoniae*, *Chlamydophila pneumoniae*, *Epstein–Barr virus*, *Herpes Simplex virus 1*, and *Cytomegalovirus* have a negative impact on sperm quality (Moretti et al. 2017). Recently, a growing interest raised on the possible impact of COVID-19, which is caused by SARS-CoV-2 virus, on fertility. Numerous possible risk factors have been found that may link SARS-CoV-2 infection with fertility problems, among them the presence of angiotensin-converting enzyme-2, the primary receptor for viral entry, in the cells of the testis (Javed et al. 2024).

From the point of view of sperm, infections play a detrimental effect on motility, DNA and acrosome integrity (Fig. 4.3). In addition, they induce damages to sperm membrane by a process named LPO. These negative effects can be mediated by toxins produced by different bacterial strains, by the direct contact between bacteria and spermatozoa, by the activation of signalling pathways related to OS, apoptosis, and inflammation (Agarwal et al. 2018) and also autophagy (Wang et al. 2021).

Epidemiological studies reported that more and more infertile men are affected by acute or chronic genitourinary tract inflammations, which often occurs without any symptoms. In particular, the negative effect on spermatogenesis seems to be due by the interplay between OS, caused by ROS increase, and proinflammatory cytokines (Azenabor et al. 2015; Paira et al. 2022).

Cytokines are a large group of proteins generated by several immune and non-immune cells in response to external stimuli, injury, or other cytokines. They participate in signal transmission between cells and perform regulatory roles in different biological processes, including cell activation, proliferation, growth, differentiation, motility but also exert modulatory effects on inflammatory reactions (Wanjari and Gopalakrishnan 2023).

The large family of cytokines includes the interleukin (IL)- 1 family, such as IL-18, the IL-6 family, the superfamily of tumor necrosis factor, IL-2, proinflammatory cytokines as IL-8, IL-12, and IL-15 (Wanjari and Gopalakrishnan 2023). Cytokines are natural components of the human semen, and at the physiological level, they act as regulators, influencing the levels of ROS in the seminal plasma. Therefore, the impact of several cytokines in the male reproductive system depends upon their concentration and pathological status, being cytokines secretion one of the first signals from the innate host defense to fight infections (Wanjari and Gopalakrishnan 2023).

In addition, the presence of leukocytes in semen samples (Fig. 4.3) can be associated with bacterial infections, but in the most cases are observed in absence of bacterial infections and inflammatory symptoms (Henkel 2024). According to the WHO guidelines (WHO 2021), the seminal leukocytes concentration should not exceed 1×10^6/mL. Leukocytes concentrations higher than this cut-off value are regarded as leukocytospermia. Leukocytospermia is generally considered an inflammatory condition that can cause sperm damage and decreased sperm parameters. However, part of the studies suggest that leukocytospermia does not necessarily indicate poor sperm functions and even a favorable role of seminal leukocytes at concentrations between 10^6 and 3×10^6/mL was suggested (Kaleli et al. 2000).

4.4 Obesity and Metabolic Syndrome

Adults with a body mass index (BMI) >30 (kg/m^2) are obese, and obesity is one of the most studied health problems in recent decades. It is estimated that 13% of the adult population worldwide is obese, but this percentage rises to more than 30% considering the adult population in the USA (Chen et al. 2017; Neto et al. 2016) and it is progressively increasing (Chooi et al. 2019).

The accumulation of white adipose tissue, hallmark of obesity, hypertension, insulin resistance, and dyslipidaemia, are mirrors of a clinical condition known as metabolic syndrome that seems to be linked to male infertility (Salvio et al. 2022). The real impact of these conditions on sperm parameters is controversial (Ameratunga et al. 2023); however, several recent studies indicate a negative association between sperm count, motility, morphology, and DNA integrity with obesity.

In a meta-analysis, Zhao and Pang (2020) reported that metabolic syndrome is associated with altered sperm parameters including DNA fragmentation concomitant with reduced levels of FSH and inhibin B.

Mounting evidence suggests that obesity negatively influences live births obtained with natural conception and ARTs (Service et al. 2023; Sciorio et al. 2024).

Interestingly, several mechanisms of action have been proposed to explain how obesity can affect sperm quality, among them hyperestrogenism, diabetes mellitus, sexual dysfunction, sperm epigenetic perturbations, hypogonadotropic hypogonadism.

One of the most accepted mechanisms is hyperestrogenism, in which serum estrogen concentration is largely increased in obese men due to an amplified peripheral conversion of testosterone to estrogens by aromatase enzyme in adipose tissue (de Boer et al. 2005) and often is concomitant with a decrease in testosterone (Stokes et al. 2015). The high systemic levels of estrogens inhibit by negative feedback the release of FSH and LH from pituitary gland, causing the testosterone hypoproduction, which, together with the decreased levels of FSH, causes alterations in the spermatogenetic process (Kahn and Brannigan 2017). This mechanism is also confirmed by the decreased levels of inhibin B in obese patients (Neto et al. 2016).

Closely associated with hormonal imbalance is the reduced level of sex hormone-binding globulin observed in obese men and patients with diabetes, which contributes to a decrease in total testosterone levels (Mah and Wittert 2010).

In addition, obesity induces systemic inflammation characterized by increased secretion of cytokines and adipocytokines (Fig. 4.3), responsible for trigger intratesticular OS (Leisegang et al. 2021) that, in turn, negatively affects semen quality. White adipose tissue is an important source of metabolic hormones such as adipocytokines such as leptin, ghrelin and obestatin, resistin, adiponectin and many others that mediate physiological homeostasis (Guerro-Millo 2004). Some of these adipocytokines induce ROS production at testicular level impairing sperm production. Other additional genetic and epigenetic factors can be involved in the pathogenesis of obesity and in the sperm production alterations (Craig et al. 2017).

4.5 Genetic Factors

The discovery of genetic causes of infertility is relevant in the era of ARTs because, in particular with intracytoplasmic sperm injection, the barriers fixed by the nature between spermatozoa and oocyte fertilization are removed. The ability to diagnose genetic factors of infertility plays a pivotal role in the counseling of the involved couples and could have implications for the health of the patient and his children. At least 15%–30% of male infertility cases are due to genetic factors that can be responsible for many idiopathic cases but also for well-known quantitative and qualitative defects of spermatogenesis, ductal dysfunction including obstructions and alterations in the hypothalamic–pituitary axis (Krausz and Riera-Escamilla 2018).

Krausz (2011) reported that the chance of carrying genetic abnormalities is highest in men with azoospermia (25%), and it gradually declines as sperm production increases.

This section aims to provide only a brief overview on the most common genetic factors affecting human spermatogenesis, highly complex process regulated by at least 2000 genes, and this number is constantly growing.

4.5.1 Y Chromosome Microdeletions

Chromosome Y is one of the most important players in controlling spermatogenesis and Y chromosome deletions/microdeletions can affect the spermatogenetic process in adulthood.

For example, deletions of the short arm of Y chromosome where sex determining region Y gene, necessary for the male sex determination, is located causes a complete sex reversal and the affected individual is phenotypically female even if they have a male genotype (McElreavy et al. 1992).

Even more interesting, from the spermatogenesis point of view, are the microdeletions of Y chromosome long arm (Tiepolo and Zuffardi 1976). Located in the euchromatic area of the long arm of the Y chromosome (Yq11), the azoospermia factor (AZF) region (Vogt et al. 1996) contains genes, highly or exclusively expressed in the testis, critical for spermatogenesis grouped as AZFa, AZFb, and AZFc. The long arm of Y chromosome harbors many ampliconic and palindromic sequences, it is prone to self-recombination during spermatogenesis, and therefore it is a sort of hotspot for intra-chromosomal deletions. Male infertility is the result of the copy number variation in Y chromosome genes caused by these deletions. The global prevalence of Yq microdeletions is 7.5% in infertile males even though a difference depending on geography and ethnicity was observed (Colaco and Modi 2018).

Krausz and Riera-Escamilla (2018) reported that most of de novo deletions likely appear during the meiotic phase of the spermatogenesis. Microdeletions occurring in any of these AZF zones are the leading molecular genetic causes of severe oligozoospermia and azoospermia; for this reason, the screening for these deletions is advisable in men with a sperm concentration lower that five million spermatozoa per ml (Krausz et al. 2014).

The AZFa group is located close the centromere of chromosome Y. The deletion analysis for the AZFa region has revealed that infertile males lacking *DEAD-Box Helicase 3 Y-Linked* (*DBY*) gene show Sertoli cell only syndrome or severe hypospermatogenesis, indicating that *DBY* is a relevant gene in the spermatogenic process (Colaco and Modi 2018). Human *DBY* expression starts during early spermatogonial proliferation (Gueler et al. 2012) this can explain the depletion of germ cells in case of deletion of the gene.

The AZFb locus is located in the central region of Yq11 and can undergo non-allelic homologous recombination with AZFc resulting in deletions.

Among different genes present in the AZFb region (Colaco and Modi 2018), two are worth of mention: *RNA-binding motif Y chromosome* (*RBMY*) and *PTP-BL related on the Y chromosome* (*PRY*) genes. *RBMY* is one of the most important genes of AZFb region and about 6 copies of this gene are dispersed within the Y chromosome. RBMY is a testis-specific splicing factor expressed in the nucleus of spermatogonia, spermatocytes, and round spermatids (Neto et al. 2016). Yan et al. (2017) demonstrated, in infertile men, a positive correlation between the number of *RBMY1* copies and sperm count and motility. *PRY* genes seem to be involved in the apoptosis regulation, leading process needed to remove abnormal sperm (Stouffs et al. 2004).

The deletion of both *RBMY* and *PRY* causes a complete spermatogenetic arrest (Ferlin et al. 2007).

Since AZFb overlaps AZFc by 1.5 Mb, combined AZFb + AZFc deletions arise and are, certainly, more frequent than isolated AZFb deletion (Neto et al. 2016).

The AZFc is located at the distal part of Yq11 and is target of 60% of all Y chromosome microdeletions in infertile men (Asero et al. 2014). The AZFa and AZFb regions are crucial in starting spermatogenetic process, whereas the AZFc region is important to complete the process of spermatogenesis (O'Flynn O'Brien et al. 2010).

Among the genes located in the AZFc group, *deleted in Azoospermia* (*DAZ*) gene, a family of four genes involved in spermatogenesis, has been the most studied (Saxena et al. 2000). The *DAZ* family genes encode highly conserved germ cell-specific RNA-binding proteins, regulate translation, and are involved in the control of meiosis and maintenance of the primordial germ cell population (Reynolds and Cooke 2005).

The most frequent deletion affecting the AZFc region is the one that removes eight genic families, including the DAZ family (Neto et al. 2016). The AZFc region is susceptible to the formation of different partial deletions; among them, the *gr/gr* deletion is the most clinically relevant, both for its frequency and for data regarding this aberration (Krausz and Riera-Escamilla 2018).

Because of the wide variability of these deletions, the clinical and histological features are variable ranging from azoospermia to oligozoospermia, depending on the extent of the deletion and the genes involved, but, in general, AZFc deletions are compatible with residual spermatogenesis (Rozen et al. 2012; Krausz and Casamonti 2017). The screening for Yq microdeletions plays a relevant role in determining the cause of male infertility, in deciding a rational strategy to manage the patients, and to correctly inform the couples about the transmission of these deletions to the male offspring born through ARTs.

4.5.2 Karyotype Anomalies and Spermatogenesis

4.5.2.1 Numerical Chromosomal Anomalies

With regard to numerical chromosomal anomalies and male infertility, the most extensive data are available for men with Klinefelter syndrome (47,XXY), whereas individuals with 47, XYY karyotype have been less thoroughly studied.

Klinefelter syndrome is associated with a typical phenotype, signs of androgen deficiency, a mild to severe hypogonadism, and a series of general health problems (Rogol 2023); here the emphasis is on the effect of this syndrome on spermatogenesis. Klinefelter syndrome is one of the leading causes of testicular failure with a frequency of 1 in 7 of patients with non-obstructive azoospermia (NOA, Punab et al. 2017). Even though the most common seminal characteristic in this syndrome is azoospermia, spermatozoa in the ejaculate and even natural pregnancy were observed, essentially referred to patients affected by mosaicism or young patients. Indeed, testicular maturation starts on time, however, around mid-puberty a surge of FSH concentration and a lesser rise of LH level impair the regular progression of spermatogenesis (Krausz and Riera-Escamilla 2018). For this reason, the age of the patient represents a predictive factor for sperm retrieval directly on the testis to perform assisted fertilization (Fainberg et al. 2019). Fluorescence in situ hybridization analysis of spermatozoa from patients affected by Klinefelter syndrome showed increased aneuploidy rates (Shi and Martin 2001), raising concerns on the safety of in vitro fertilization for these patients.

Subjects with 47,XYY karyotype have much less testicular dysfunction than those with Klinefelter syndrome. Sperm concentration range from normozoospermia to azoospermia and the increased rate of gonosomal aneuploidies in spermatozoa and pre-implantation embryos are observed (Gonzalez-Merino et al. 2007).

4.5.2.2 Structural Chromosomal Anomalies

Besides numerical abnormalities, structural chromosomal anomalies are 5–10 times more frequent in infertile men (Krausz and Riera-Escamilla 2018). The formation of normal bivalents during meiosis is disrupted in patients with structural abnormalities (mainly with respect to translocations), leading to impaired meiosis and maturation arrest of spermatogenesis. In most of the patients with structural chromosome abnormalities, oligozoospermia is observed. The frequency of reciprocal translocations, Robertsonian translocations, and inversions is higher in men with oligozoospermia compared with azoospermic men and men in the general population (Harton and Tempest 2012).

In reciprocal translocation, the exchange of chromosomal segments between two non-homologous chromosomes establishes new linkage groups. Robertsonian translocations are characterized by centric fusion of two acrocentric chromosomes: the long arms are combined to form one large chromosome and one small chromosome (lost without any consequence on viability), the karyotype of the carrier counts 45 chromosomes. Among infertile men, the Robertsonian translocation t (13;14) is very common. Cytogenetic studies on infertile men and couples experiencing recurrent pregnancy loss (RPL) revealed that, in comparison to the proportion reported in newborns, the percentage of t (13;14) carriers was two to 25 times greater in infertile males and four to 17 times higher in couples experiencing RPL (Tunç and Ilgaz 2022).

Recently, Chen and Zhou (2022) performed a systematic review on the relationship between semen parameters and the presence of autosomal reciprocal translocations or Robertsonian translocations. They concluded that compared with controls, autosomal reciprocal translocation and Robertsonian translocation carriers showed lower sperm concentration, total motility, progressive motility, and normal morphology.

It is well known that both these chromosomal structural alterations can affect meiotic chromosomal segregation with the result of an increase of aneuploid spermatozoa (Anton et al. 2011; Baccetti et al. 2005; Moretti et al. 2009b; Shi and Martin 2001). Moreover, in patients with chromosomal translocations, the particular meiotic configuration of translocated chromosomes can interfere with the correct meiotic behavior of the other chromosomes; this phenomenon is known as the interchromosomal effect.

The case of balanced X-autosome translocations is peculiar, because male carriers of this kind of translocation are almost invariably infertile, affected by azoospermia or severe oligozoospermia due to a spermatogenetic arrest (Van Assche et al. 1996).

Chromosomal inversions are structural rearrangements described in almost all human chromosomes. They occur when in a chromosome, a segment breaks off and re-attaches within the same chromosome, but in inverted orientation; the re-oriented segment can contain the centromere (pericentric) or not (paracentric). In a recent review, it was reported that the type of inversion and sperm parameters of male chromosomal inversion carriers did not affect the ploidy status of embryos (Tong et al. 2022). Among the different chromosomal inversions, pericentric inversion of chromosome 9 represents a common structural polymorphic variant in human population; however, its impact on reproductive outcomes is still debated. Liu et al. (2024) recently demonstrated that the pericentric inversion of chromosome 9 does not influence sperm parameters except for a slight decrease of progressive motility and does not increase reproductive risks overall.

4.5.2.3 Monogenic Causes Affecting Spermatogenetic Process

Discussing on gene mutations affecting the spermatogenesis, we can refer to the systematic review of Houston et al. (2021) which reports a considerable number of novel gene–disease relationships in the field of male infertility. In this review, the authors identified 120 genes that were moderately, strongly, or definitively linked to 104 infertility phenotypes, which represent a 33% increase from the number identified in 2019. Endocrine diseases or reproductive system syndromes (n = 67) and isolated infertility phenotypes (n = 36) accounted for the bulk of confident gene-disease relationships, but a minority (n = 17) were connected to syndromic infertility, including primary ciliary dyskinesia. These genes are related to hypothalamus development/function, pituitary gland dysfunction, adrenal gland dysfunction, abnormal development of urethra and vas deferens, abnormal development of reproductive organs, Leydig cell dysfunction, Sertoli cell only syndrome, pre-meiotic arrest, meiotic/germ cell arrest, spermiogenesis defects, fertilization defect (Houston et al. 2021).

Here, we focus on several common mutations that impact on spermatogenetic process.

For example, NOA is generally due to congenital factors leading to spermatogenic failure. Several chromosome abnormalities have been identified, but the knowledge of the monogenic causes of NOA is very limited. The 38 NOA genes that have been identified and described recently serve as crucial guardians of meiosis and transcriptional and endocrine regulators of reproduction (Kasak and Laan 2021).

In case of quantitative spermatogenesis disturbances, androgen receptor (AR) gene mutations can be supposed. *AR* gene is located the X chromosome and more than one thousand mutations have been discovered (Gottlieb et al. 2012). The phenotype ranges from complete insensitivity to androgen (Morris syndrome), through a partial androgen insensitivity syndrome, to mild androgen insensitivity syndrome with a normal male genital phenotype and impaired sperm production (Krausz and Riera-Escamilla 2018).

Congenital hypogonadotropic hypogonadism and Kallmann syndrome, in which anosmia is present, are both conditions that prevent normal pubertal development and cause infertility in affected men and women. Since residual spermatogenesis may be present in some cases and hormonal treatments are available, affected individuals have the possibility to procreate through ARTs. These two diseases are heterogeneous both from phenotype and genotype point of view and the risk of transmission largely depends on the gene involved (Maione et al. 2018).

In certain cases, the oligozoospermia or azoospermia depend on the ductal obstruction or dysfunction, meaning an interference with the sperm transport system that can occur at different levels in the genitourinary tract. Generally, the normal testis histology and therefore a normal spermatogenesis is present (Krausz and Riera-Escamilla 2018). CAVD with intact renal structure is the only congenital variant with a recognized genetic etiology. CAVD is caused by recessive mutations of the gene *cystic fibrosis transmembrane conductance regulator-CFTR-* that, if mutated, causes cystic fibrosis. CAVD may show several clinical aspects depending on whether it is bilateral, unilateral, complete or partial, and associated or not associated with other abnormalities of the male genitourinary tract (Bieth et al. 2021).

In the next chapter, gene mutations that cause peculiar sperm morphological alterations (systematic sperm defects) will be discussed.

4.6 Pollutants and Lifestyle

Numerous environmental factors to which humans are exposed are harmful for male reproductive capacity. Today, the suspected increased incidence of male infertility induced by such factors are of great concern and the amount of available data is significant. Far from providing a detailed assessment on this topic, the purpose of this paragraph is to give a quick general view of the most important toxicants or lifestyle habits that are reputed to exert undesirable consequence on male infertility.

A recent meta-analysis from a Chinese group showed that air pollution impairs male infertility affecting semen quality. In particular, $PM_{2.5}$ (particulate matter with a diameter of 2.5 µ or less), PM_{10} (particulate matter with a diameter of 10 µ or less), NO_2, and SO_2 exposure was related to a decrease of sperm concentration and motility (Liu et al. 2023).

A class of the most studied compounds are endocrine disruptors. These are exogenous substances, which alter functions of the endocrine system, causing adverse health effects and compromising fertility. These endocrine disruptors can have estrogenic or anti-androgenic mechanism of action, obviously altering the normal physiology. Just to mention some of them, bisphenol A (Kortenkamp et al. 2022), parabens, polychlorinated biphenyls, polybrominated diphenyl ethers, phthalates, ultraviolet filters, and some pesticides (dichlorodiphenyltrichloroethane) and others can act as xenoestrogens (Cofone et al. 2023; Rodprasert et al. 2023). As a whole, these studies demonstrate an association between the environmental and/or occupational exposure to these toxicants and the alteration of some sperm parameters as sperm concentration, motility, morphology, and vitality (Cannarella et al. 2023; Rodprasert et al. 2023).

Other than classical semen parameters, DNA integrity can also be affected by pollutant exposure (Muratori et al. 2019) and it is well known that high levels of sperm DNA fragmentation have been associated with adverse reproductive outcomes, including low natural and assisted pregnancy rates, as well as altered embryo development, and RPL (Marinaro 2023).

When discussing endocrine disruptors, it is important to mention microplastics and nanoplastics, which originate from the degradation of plastic waste. The topic is recent but several studies, in particular performed in animal models, revealed a deleterious effect of microplastic exposure in male reproduction and sperm quality (D'Angelo and Meccariello 2021). In fact, microplastics and nanoplastics can absorb harmful chemicals such as endocrine disruptors, but also heavy metals, and they can act as a transport medium for them that accumulate in the different organisms.

In this regard, it should be noted that smaller particles exhibit enhanced absorption capacity and a larger surface area can facilitate the release of greater amounts of toxicants and endocrine disruptors (Ullah et al. 2023).

Programs for human biomonitoring might be considered as an effective tool for some surveillance initiatives aiming to identify and minimize the risk exposure via early identification of people who are more exposed.

Three of the most common unhealthy habits that can affect sperm quality and fertility are alcohol, tobacco, and drug addictions. Alcohol consumption is often considered socially acceptable, but its negative effects on gonadal function have been consistently reported in the last 30 years.

Finelli et al. (2021) summarized in a narrative review the relation between chronic and excessive alcohol consumption and altered hormonal homeostasis, semen quality, and genetic and epigenetic regulations, due to the development of OS that, as reported above, damage spermatozoa. Recently, a meta-analysis suggested that chronic alcohol consumption inhibits the gonadal axis in healthy men, decreasing testosterone and sex hormone-binding globulin levels and increasing estradiol concentration (Santi et al. 2024).

Maternal smoking, as well as fetal exposure to paternal smoking during pregnancy (Haervig et al. 2020), has been associated with reduced semen quality in male offspring. However, several conflicting results have been reported in the literature (Rodprasert et al. 2023; Sansone et al. 2018).

Various meta-analyzes have indicated that smoking is a risk factor for impaired semen quality in men, particularly affecting sperm concentration, motility, and the percentage of sperm with normal morphology. Furthermore, smoking has been identified as the lifestyle factor with the most detrimental effect on DNA integrity, leading to fragmentation (Szabó et al. 2023).

Cigarette smoke contains many chemicals, like nicotine, carbon monoxide, and polycyclic aromatic hydrocarbons, lead and cadmium that can trigger ROS production, apoptosis, cause germ cell genome and epigenome alterations.

Cannabis is the most commonly used drug in the world, especially by people in reproductive age. Cannabis plays biological effect through the endocannabinoid system, whose receptors are present in both female and male reproductive system (Lo et al. 2022).

As regards male infertility, cannabis consumption plays a negative effect on sperm parameters, and on sperm nuclear quality by increasing chromosomal abnormalities, DNA fragmentation (Verhaeghe et al. 2020) but also a change to the sperm DNA methylation, thus influencing the sperm epigenome (Schrott and Murphy 2020).

Growing evidence on the importance of nutrition in reproduction is emerging, even though the biochemical mechanisms involved in the modulation of sperm quality are not completely clarified (Ferramosca and Zara 2022). The Mediterranean diet seems to be protective against male infertility. In fact, dietary patterns with the strongest evidence for associations with improved sperm quality are those composed by consistent use of vegetables, fruits, whole grains, nuts, low-fat dairy, and seafood, and lower intakes of sweets, sugar-sweetened beverages, and red and processed meats (LaPointe et al. 2023; Pecora et al. 2023).

In particular, a diet rich in saturated fatty acids, with reduced PUFAs and with an unbalanced omega-6/omega-3 PUFA ratio negatively affect sperm parameters (Collodel et al. 2020; Ferramosca and Zara 2022).

In addition, increased intake of selenium and zinc, coenzymeQ10 and carnitine supplements have been positively related to sperm quality (Salas-Huetos et al. 2019).

Finally, male reproductive health can also be affected by other factors. For example, physiological stress (Ilacqua et al. 2018), paternal aging (Collodel et al. 2021; du Fossé et al. 2020; Kaltsas et al. 2023), and also mobile telephones (Gameiro et al. 2016) can play a negative influence on sperm parameters and DNA quality.

4.7 Focus on Epigenetics, Spermatogenesis, and Spermatozoa

Epigenetic is a sort of system to transfer information about the lived experiences to offspring (Greeson et al. 2023). Epigenetics is the study of several processes that alter gene expression without changing the DNA sequence; some of these processes include DNA methylation, histone modifications, and chromatin remodeling, non-coding RNA molecules. Essentially, the chromatin can be in a sort of permissive condition that allows transcription and gene expression or in a repressive condition inhibiting the gene expression. The genome is widely remodeled from the epigenetic point of view during mammalian germ cell development in both male and female germlines, and this phenomenon is crucial for imprinting and reprogramming in early embryos. The organization of the chromatin is different in spermatozoa and oocytes. Sperm DNA is highly methylated and tightly packaged (histones are mostly replaced with protamines), whereas oocytes have prevalently an open chromatin (McSwiggin and O'Doherty 2018; Muratori and De Geyter 2019). Spermatozoa have increased chance of accumulating epigenetic errors because the constant mitotic and meiotic cycles that characterize the spermatogenetic process.

DNA methylation consists in the addition of a methyl radical to nitrogenous base cytosine of cytosine-guanine dinucleotides (CpG islands) by DNA methyltransferases: hyper-methylation of CpG islands is linked with gene silencing, while hypomethylation to gene expression.

Various environmental and lifestyle factors previously described are observed to alter sperm epigenetics and to regulate gene expression during early embryonic development. The relationship between DNA methylation, infertility and abnormal semen parameters is not completely clear. Oligozoospermia seems to be associated with anomalies of DNA methylation of imprinted regions. In a systematic review, Åsenius et al. (2020) concluded that sperm DNA methylation is influenced by several factors but in particular a pivotal role is played by cigarette smoking, aging (Ashapkin et al. 2023,) and environmental pollutants. Growing evidence that paternal obesity (Craig et al. 2017; Keyhan et al. 2021) influence offspring metabolic and reproductive phenotypes by means of epigenetic reprogramming are available in the literature.

During the process of spermiogenesis, the round spermatids are modeled in spermatozoa. One of the most relevant events is related to histone-to-protamine transition (see Sect. 3.2.2) that enables the tight packaging of sperm DNA that guarantees the acquisition of a hydrodynamic nuclear shape and the protection of DNA from damages. Post-translational modifications of histone tails include methylation, phosphorylation, ubiquitination, SUMOylation, and acetylation, which in different combinations can regulate gene expression (Ghai and Kader 2022).

Data on this topic came from studies on animal models; research on the relationship between sperm chromatin structure and fertility are scant (McSwiggin and O'Doherty 2018). Defects on histone-protamine replacement can cause a decrease of sperm chromatin condensation with consequent negative implication in male fertility (Francis et al. 2014). Indeed, the reduction or premature histone hyperacetylation provokes abnormal spermatogenesis and the excessive histone retention is frequently observed in infertile men (Zhang et al. 2006).

Spermatozoa show low transcription levels, but growing evidence demonstrates that sperm carry a wide variety of RNA molecules, both coding and non-coding, required for spermatogenesis, early development, and epigenetic inheritance. In particular, non-coding RNAs loaded in human sperm provide evidence of being reliable carriers for transgenerational inheritance. The non-coding RNA transcriptome is composed by two main groups of RNAs based their length: short non-coding RNAs, and long non-coding RNAs. The most important groups of non-coding RNA are sperm P-element induced wimpy testis (Piwi)-RNAs (piRNAs) and microRNAs (miRNAs) that play an essential role in spermatogenesis and are involved in male fertility disorders. The piRNAs are small regulatory RNAs of 23 to 31 nucleotides in length, predominantly expressed in the male germline. They are essential in preserving the integrity of the germline genome protecting it from "injuries" caused by harmful selfish nucleic acid elements, such as transposable elements (Olotu et al. 2023). In addition, growing evidence suggests that PIWI-piRNA complex plays a role in the posttranscriptional regulation of non-coding and protein-coding genes (Wang et al. 2023).

The miRNAs are an important group of molecules of single-stranded non-coding nucleotides (20–23 nucleotides), and one of the most represented ribonucleoprotein complexes in the cell. The miRNA mechanisms of action consist in the stopping or destruction of targets mRNAs, translation repression, and other indirect pathways to inhibit protein synthesis. In this way, miRNAs regulated gene expression in a cell-specific manner during the spermatogenetic process, controlling each step of male germ cell differentiation (Walker 2022). Therefore, their alterations can lead to male infertility (Burgos et al. 2022; Joshi et al. 2023; Salas-Huetos et al. 2014). The miRNAs are present in the sperm cells, probably originated from the testis and during epididymal maturation, and in seminal plasma, originated from the testis, epididymis, and accessory glands, and both have been explored as potential biomarker for male factor fertility.

RNA classes involved in epigenetic transmission are expected to yield significant and potentially surprising insights in the coming years.

Just consider that recently, Tomar et al. (2024) discovered that paternal health at the time of conception is important for offspring metabolism; in particular, they found that mitochondrial tRNAs are diet-induced and sperm-borne, demonstrating father-to-offspring transfer of sperm mitochondrial RNAs at fertilization.

4.8 Conclusion

Several anatomical, genetics, environmental, and lifestyle factors can play a negative impact on the spermatogenetic process affecting male infertility, which, in many cases, remains undiagnosed. Therefore, it is necessary to approach the different potential causes of male infertility, studying new key factors and diagnostic biomarkers. In the last few decades, OS has emerged as a common denominator of many conditions that impair sperm quality and, at the same time, the identification of small non-coding RNAs and epigenetics play a relevant role in understanding the underlying mechanism of the diseases. Obviously, male infertility is not always preventable, but the knowledge of the effects of pathologies and other conditions on human spermatogenesis is needed in order to plan appropriate strategies for the management of infertile patients.

Acknowledgments We would like to express our extreme gratitude to Dr. Laura Liguori for her support in creating images used in this chapter.

References

Agarwal A, Mulgund A, Hamada A, Chyatte MR (2015) A unique view on male infertility around the globe. Reprod Biol Endocrinol 13:37. https://doi.org/10.1186/s12958-015-0032-1

Agarwal A, Rana M, Qiu E, AlBunni H, Bui AD, Henkel R (2018) Role of oxidative stress, infection and inflammation in male infertility. Andrologia 50(11):e13126. https://doi.org/10.1111/and.13126

Agarwal A, Parekh N, Panner Selvam M, Henkel R, Shah R, Homa S, Harlev A (2019) Male oxidative stress infertility (MOSI): proposed terminology and clinical practice guidelines for management of idiopathic male infertility. WJMH 37(3):296–312. https://doi.org/10.5534/wjmh.190055

Aitken RJ, Baker MA (2013) Oxidative stress, spermatozoa and leukocytic infiltration: relationships forged by the opposing forces of microbial invasion and the search for perfection. J Reprod Immunol 100(1):11–19. https://doi.org/10.1016/j.jri.2013.06.005

Aitken RJ, Smith TB, Jobling MS, Baker MA, De Iuliis GN (2014) Oxidative stress and male reproductive health. Asian J Androl 16(1):31–38. https://doi.org/10.4103/1008-682X.122203

Aitken RJ, Drevet JR, Moazamian A, Gharagozloo P (2022) Male infertility and oxidative stress: a focus on the underlying mechanisms. Antioxidants 11(2):306. https://doi.org/10.3390/antiox11020306

Ameratunga D, Gebeh A, Amoako A (2023) Obesity and male infertility. Best Pract Res Clin Obstet Gynaecol 90:102393. https://doi.org/10.1016/j.bpobgyn.2023.102393

Anton E, Vidal F, Blanco J (2011) Interchromosomal effect analyses by sperm FISH: incidence and distribution among reorganization carriers. Syst Biol Reprod Med 57(6):268–278. https://doi.org/10.3109/19396368.2011.633682

Åsenius F, Danson AF, Marzi SJ (2020) DNA methylation in human sperm: a systematic review. Hum Reprod Update 26(6):841–873. https://doi.org/10.1093/humupd/dmaa025

Asero P, Calogero AE, Condorelli RA, Mongioi' L, Vicari E, Lanzafame F, Crisci R, La Vignera S (2014) Relevance of genetic investigation in male infertility. J Endocrinol Investig 37(5):415–427. https://doi.org/10.1007/s40618-014-0053-1

Ashapkin V, Suvorov A, Pilsner JR, Krawetz SA, Sergeyev O (2023) Age-associated epigenetic changes in mammalian sperm: implications for offspring health and development. Hum Reprod Update 29(1):24–44. https://doi.org/10.1093/humupd/dmac033

Azenabor A, Ekun AO, Akinloye O (2015) Impact of inflammation on male reproductive tract. J Reprod Infertil 16(3):123–129

Baccetti B, Collodel G, Marzella R, Moretti E, Piomboni P, Scapigliati G, Serafini F (2005) Ultrastructural studies of spermatozoa from infertile males with Robertsonian translocations and 18, X. Y aneuploidies Hum Reprod 20(8):2295–2300. https://doi.org/10.1093/humrep/dei050

Baskaran S, Finelli R, Agarwal A, Henkel R (2021) Diagnostic value of routine semen analysis in clinical andrology. Andrologia 53(2):e13614. https://doi.org/10.1111/and.13614

Bhattacharya I, Sharma SS, Majumdar SS (2024) Etiology of male infertility: an update. Reprod Sci 31(4):942–965. https://doi.org/10.1007/s43032-023-01401-x
Bieth E, Hamdi SM, Mieusset R (2021) Genetics of the congenital absence of the vas deferens. Hum Genet 140(1):59–76. https://doi.org/10.1007/s00439-020-02122-w
Burgos CF, Cikutovic R, Alarcón M (2022) MicroRNA expression in male infertility. Reprod Fertil Dev 34(12):805–818. https://doi.org/10.1071/RD21131
Cannarella R, Gül M, Rambhatla A, Agarwal A (2023) Temporal decline of sperm concentration: role of endocrine disruptors. Endocrine 79(1):1–16. https://doi.org/10.1007/s12020-022-03136-2
Cannarella R, Shah R, Saleh R, Boitrelle F, Hamoda TAA, Singh R, Salvio G, Toprak T, Falcone M, Gul M, Dimitriadis F, Rambhatla A, Russo GI, Ko E, Zini A, Kavoussi P, Phuoc NHV, Kandil H, Ghayda RA, Birowo P, Gherabi N, Ceyhan E, Dong J, Malhotra V, Durairajanayagam D, Kolbasi B, Bahar F, Calik G, Çayan S, Pinggera GM, Calogero AE, Rajmil O, Mostafa T, Atmoko W, Harraz AM, Le TV, de la Rosette J, Hakim L, Pescatori E, Sergeyev O, Rashed A, Saini P, Agarwal A (2024) Effects of varicocele repair on sperm DNA fragmentation and seminal malondialdehyde levels in infertile men with clinical varicocele: a systematic review and meta-analysis. World J Mens Health 42(2):321–337. https://doi.org/10.5534/wjmh.230235
Chen X, Zhou C (2022) Reciprocal translocation and Robertsonian translocation in relation to semen parameters: a retrospective study and systematic review. Andrologia 54(1):e14262. https://doi.org/10.1111/and.14262
Chen H, Mruk D, Xiao X, Cheng CY (2017) Human spermatogenesis and its regulation. In: Winters S, Huhtaniemi I (eds) Male hypogonadism. Contemporary endocrinology. Humana Press, Cham. https://doi.org/10.1007/978-3-319-53298-1_3
Cheng L, Sun P, Xie X, Sun D, Zhou Q, Yang S, Xie Q, Zhou X (2000) Hepatitis B virus surface protein induces oxidative stress by increasing peroxides and inhibiting antioxidant defences in human spermatozoa. Reprod Fertil Dev 32(14):1180–1189. https://doi.org/10.1071/RD20130
Chooi YC, Ding C, Magkos F (2019) The epidemiology of obesity. Metabolism 92:6–10
Clavijo RI, Carrasquillo R, Ramasamy R (2017) Varicoceles: prevalence and pathogenesis in adult men. Fertil Steril 108(3):364–369. https://doi.org/10.1016/j.fertnstert.2017.06.036
Cofone L, Pindinello I, D'Ancona G, Grassi F, Antonucci A, Vitali M, Protano C (2023) Human semen quality and environmental and occupational exposure to pollutants: a systematic review. Ann Ig 35(6):660–669. https://doi.org/10.7416/ai.2023.2581
Colaco S, Modi D (2018) Genetics of the human Y chromosome and its association with male infertility. Reprod Biol Endocrinol 16(1):14. https://doi.org/10.1186/s12958-018-0330-5
Collodel G, Moretti E, Micheli L, Menchiari A, Moltoni L, Cerretani D (2015) Semen characteristics and malondialdehyde levels in men with different reproductive problems. Andrology 3(2):280–286. https://doi.org/10.1111/andr.297
Collodel G, Moretti E, Longini M, Pascarelli NA, Signorini C (2018) Increased F2-Isoprostane levels in semen and immunolocalization of the 8-iso prostaglandin F2α in spermatozoa from infertile patients with varicocele. Oxidative Med Cell Longev 2018:7508014. https://doi.org/10.1155/2018/7508014
Collodel G, Castellini C, Lee JC, Signorini C (2020) Relevance of fatty acids to sperm maturation and quality. Oxidative Med Cell Longev 2020:7038124. https://doi.org/10.1155/2020/7038124
Collodel G, Ferretti F, Masini M, Gualtieri G, Moretti E (2021) Influence of age on sperm characteristics evaluated by light and electron microscopies Sci Rep 11(1):4989. https://doi.org/10.1038/s41598-021-84051-w
Collodel G, Moretti E, Noto D, Corsaro R, Signorini C (2022) Oxidation of polyunsaturated fatty acids as a promising area of research in infertility. Antioxidants (Basel) 11(5):1002. https://doi.org/10.3390/antiox11051002
Craig JR, Jenkins TG, Carrell DT, Hotaling JM (2017) Obesity, male infertility, and the sperm epigenome. Fertil Steril 107(4):848–859. https://doi.org/10.1016/j.fertnstert.2017.02.115
Damsgaard J, Joensen UN, Carlsen E, Erenpreiss J, Blomberg Jensen M, Matulevicius V, Zilaitiene B, Olesen IA, Perheentupa A, Punab M, Salzbrunn A, Toppari J, Virtanen HE, Juul A, Skakkebæk NE, Jørgensen N (2016) Varicocele is associated with impaired semen quality

and reproductive hormone levels: a study of 7035 healthy Young men from six European countries. Eur Urol 70(6):1019–1029. https://doi.org/10.1016/j.eururo.2016.06.044
D'Angelo S, Meccariello R (2021) Microplastics: a threat for male fertility. Int J Environ Res Public Health 18(5):2392. https://doi.org/10.3390/ijerph18052392
Datta J, Palmer MJ, Tanton C, Gibson LJ, Jones KG, Macdowall W, Glasier A, Sonnenberg P, Field N, Mercer CH, Johnson AM, Wellings K (2016) Prevalence of infertility and help seeking among 15000 women and men. Hum Reprod 31(9):2108–2118. https://doi.org/10.1093/humrep/dew123
de Boer H, Verschoor L, Ruinemans-Koerts J, Jansen M (2005) Letrozole normalizes serum testosterone in severely obese men with hypogonadotropic hypogonadism. Diabetes Obesity Metabol 7:211–215
De Jonge C (2017) Biological basis for human capacitation-revisited. Hum Reprod Update 23(3):289–299. https://doi.org/10.1093/humupd/dmw048
Dennis EA, Norris PC (2015) Eicosanoid storm in infection and inflammation. Nat Rev Immunol 15(8):511–523. https://doi.org/10.1038/nri3859
Drevet JR, Aitken RJ (2019) Oxidative damage to sperm DNA: attack and Defense. Adv Exp Med Biol 1166:107–117. https://doi.org/10.1007/978-3-030-21664-1_7
Drevet JR, Hallak J, Nasr-Esfahani MH, Aitken RJ (2022) Reactive oxygen species and their consequences on the structure and function of mammalian spermatozoa. Antioxid Redox Signal 37(7–9):481–500. https://doi.org/10.1089/ars.2021.0235
du Fossé NA, van der Hoorn MP, van Lith JMM, le Cessie S, Lashley EELO (2020) Advanced paternal age is associated with an increased risk of spontaneous miscarriage: a systematic review and meta-analysis. Hum Reprod Update 26(5):650–669. https://doi.org/10.1093/humupd/dmaa010
Eisenberg ML, Esteves SC, Lamb DJ, Hotaling JM, Giwercman A, Hwang K, Cheng YS (2023) Male infertility Nat Rev Dis Primers 9(1):49. https://doi.org/10.1038/s41572-023-00459-w
Esteves SC, Zini A, Coward RM, Evenson DP, Gosálvez J, Lewis SEM, Sharma R, Humaidan P (2021) Sperm DNA fragmentation testing: summary evidence and clinical practice recommendations. Andrologia 53(2):e13874. https://doi.org/10.1111/and.13874
Fainberg J, Kashanian JA (2019) Recent advances in understanding and managing male infertility. F1000 Res 8:F1000 faculty Rev-670. https://doi.org/10.12688/f1000research.17076.1
Fainberg J, Hayden RP, Schlegel PN (2019) Fertility management of Klinefelter syndrome. Expert Rev Endocrinol Metab 14(6):369–380. https://doi.org/10.1080/17446651.2019.1671821
Farsimadan M, Motamedifar M (2020) Bacterial infection of the male reproductive system causing infertility. J Reprod Immunol 142:103183. https://doi.org/10.1016/j.jri.2020.103183
Ferlin A, Raicu F, Gatta V, Zuccarello D, Palka G, Foresta C (2007) Male infertility: role of genetic background. Reprod Biomed Online 14:734–745
Ferramosca A, Zara V (2022) Diet and male fertility: the impact of nutrients and antioxidants on sperm energetic metabolism. Int J Mol Sci 23(5):2542. https://doi.org/10.3390/ijms23052542
Finelli R, Mottola F, Agarwal A (2021) Impact of alcohol consumption on male fertility potential: a narrative review. Int J Environ Res Public Health 19(1):328. https://doi.org/10.3390/ijerph19010328
Francis S, Yelumalai S, Jones C, Coward K (2014) Aberrant protamine content in sperm and consequential implications for infertility treatment. Hum Fertil (Camb) 17(2):80–99. https://doi.org/10.3109/14647273.2014.915347
Gameiro S, Boivin J, Dancet E, Emery M, Thorn P, Van den Broeck U et al (2016) Qualitative research in the ESHRE guideline 'routine psychosocial care in infertility and medically assisted reproduction—a guide for staff guideline development group of the ESHRE guideline on psychosocial Care in Infertility and Medically Assisted Reproduction. Hum Reprod 31:1928–1929
Garolla A, Graziani A, Grande G, Ortolani C, Ferlin A (2024) HPV-related diseases in male patients: an underestimated conundrum. J Endocrinol Investig 47(2):261–274. https://doi.org/10.1007/s40618-023-02192-3

Gat Y, Zukerman Z, Chakraborty J, Gornish M (2005) Varicocele, hypoxia and male infertility. Fluid mechanics analysis of the impaired testicular venous drainage system. Hum Reprod 20(9):2614–2619. https://doi.org/10.1093/humrep/dei089
Ghai M, Kader F (2022) A review on epigenetic inheritance of experiences in humans. Biochem Genet 60(4):1107–1140. https://doi.org/10.1007/s10528-021-10155-7
Gonzalez-Merino E, Hans C, Abramowicz M, Englert Y, Emiliani S (2007) Aneuploidy study in sperm and preimplantation embryos from nonmosaic 47. XYY men Fertil Steril 88(3):600–606. https://doi.org/10.1016/j.fertnstert.2006.12.020
Gottlieb B, Beitel LK, Nadarajah A, Paliouras M, Trifiro M (2012) The androgen receptor gene mutations database: 2012 update. Hum Mutat 33(5):887–894. https://doi.org/10.1002/humu.22046
Greeson KW, Crow KMS, Edenfield RC, Easley CA 4th (2023) Inheritance of paternal lifestyles and exposures through sperm DNA methylation. Nat Rev Urol 20(6):356–370. https://doi.org/10.1038/s41585-022-00708-9
Gueler B, Sonne SB, Zimmer J, Hilscher B, Hilscher W, Græm N, Rajpert-De Meyts E, Vogt PH (2012) AZFa protein DDX3Y is differentially expressed in human male germ cells during development and in testicular tumours: new evidence for phenotypic plasticity of germ cells. Hum Reprod 27:1547–1555
Guerro-Millo M (2004) Adipose tissue and adipokines: for better or worse. Diabetes Metab 30:13–19
Haervig KK, Høyer BB, Giwercman A, Hougaard KS, Ramlau-Hansen CH, Specht IO, Toft G, Bonde JP, Søgaard Tøttenborg S (2020) Fetal exposure to paternal smoking and semen quality in the adult son. Andrology 8(5):1117–1125. https://doi.org/10.1111/andr.12782
Harton GL, Tempest HG (2012) Chromosomal disorders and male infertility. Asian J Androl 14(1):32–39. https://doi.org/10.1038/aja.2011.66
Hasan H, Bhushan S, Fijak M, Meinhardt A (2022) Mechanism of inflammatory associated impairment of sperm function. Spermatogenesis and Steroidogenesis Front Endocrinol (Lausanne) 13:897029. https://doi.org/10.3389/fendo.2022.897029
Henkel R (2024) Leukocytospermia and/or Bacteriospermia: impact on male infertility. J Clin Med 13(10):2841. https://doi.org/10.3390/jcm13102841
Houston BJ, Riera-Escamilla A, Wyrwoll MJ, Salas-Huetos A, Xavier MJ, Nagirnaja L, Friedrich C, Conrad DF, Aston KI, Krausz C, Tüttelmann F, O'Bryan MK, Veltman JA, Oud MS (2021) A systematic review of the validated monogenic causes of human male infertility: 2020 update and a discussion of emerging gene-disease relationships. Hum Reprod Update 28(1):15–29. https://doi.org/10.1093/humupd/dmab030
Ilacqua A, Izzo G, Emerenziani GP, Baldari C, Aversa A (2018) Lifestyle and fertility: the influence of stress and quality of life on male fertility. Reprod Biol Endocrinol 16(1):115. https://doi.org/10.1186/s12958-018-0436-9
Javed A, Ramaiah MK, Saba K (2024) Exploring spermatogenesis post-SARS-CoV-2: a comprehensive review of male reproductive health. J Men's Health 20(3):14–24
Jensen CFS, Østergren P, Dupree JM, Ohl DA, Sønksen J, Fode M (2017) Varicocele and male infertility. Nat Rev Urol 14(9):523–533. https://doi.org/10.1038/nrurol.2017.98
Jing J, Peng Y, Fan W, Han S, Peng Q, Xue C, Qin X, Liu Y, Ding Z (2023) Obesity-induced oxidative stress and mitochondrial dysfunction negatively affect sperm quality. FEBS Open Bio 13(4):763–778. https://doi.org/10.1002/2211-5463.13589
Joshi M, Sethi S, Mehta P, Kumari A, Rajender S (2023) Small RNAs, spermatogenesis, and male infertility: a decade of retrospect. Reprod Biol Endocrinol 21(1):106. https://doi.org/10.1186/s12958-023-01155-w
Kahn BE, Brannigan RE (2017) Obesity and male infertility. Curr Opin Urol 27(5):441–445. https://doi.org/10.1097/MOU.0000000000000417
Kaleli S, Oçer F, Irez T, Budak E, Aksu MF (2000) Does leukocytospermia associate with poor semen parameters and sperm functions in male infertility? The role of different seminal leukocyte concentrations. Eur J Obstet Gynecol Reprod Biol 89(2):185–191. https://doi.org/10.1016/s0301-2115(99)00204-3

Kaltsas A, Moustakli E, Zikopoulos A, Georgiou I, Dimitriadis F, Symeonidis EN, Markou E, Michaelidis TM, Tien DMB, Giannakis I, Ioannidou EM, Papatsoris A, Tsounapi P, Takenaka A, Sofikitis N, Zachariou A (2023) Impact of advanced paternal age on fertility and risks of genetic disorders in offspring. Genes (Basel) 14(2):486. https://doi.org/10.3390/genes14020486

Kasak L, Laan M (2021) Monogenic causes of non-obstructive azoospermia: challenges, established knowledge, limitations and perspectives. Hum Genet 140(1):135–154. https://doi.org/10.1007/s00439-020-02112-y

Keyhan S, Burke E, Schrott R, Huang Z, Grenier C, Price T, Raburn D, Corcoran DL, Soubry A, Hoyo C, Murphy SK (2021) Male obesity impacts DNA methylation reprogramming in sperm. Clin Epigenetics 13(1):17. https://doi.org/10.1186/s13148-020-00997-0

Kirby EW, Wiener LE, Rajanahally S, Crowell K, Coward RM (2016) Undergoing varicocele repair before assisted reproduction improves pregnancy rate and live birth rate in azoospermic and oligospermic men with a varicocele: a systematic review and meta-analysis. Fertil Steril 106(6):1338–1343. https://doi.org/10.1016/j.fertnstert.2016.07.1093

Koksal IT, Ishak Y, Usta M, Danisman A, Guntekin E, Bassorgun IC, Ciftcioglu A (2007) Varicocele-induced testicular dysfunction may be associated withdisruption of blood-testis barrier. Arch Androl 53(2007):43–48

Kortenkamp A, Martin O, Ermler S, Baig A, Scholze M (2022) Bisphenol a and declining semen quality: a systematic review to support the derivation of a reference dose for mixture risk assessments. Int J Hyg Environ Health 241:113942. https://doi.org/10.1016/j.ijheh.2022.113942

Krausz C (2011) Male infertility: pathogenesis and clinical diagnosis. Best Pract Res Clin Endocrinol Metab 25:271–285

Krausz C, Casamonti E (2017) Spermatogenic failure and the Y chromosome. Hum Genet 136(5):637–655. https://doi.org/10.1007/s00439-017-1793-8

Krausz C, Riera-Escamilla A (2018) Genetics of male infertility. Nat Rev Urol 15(6):369–384. https://doi.org/10.1038/s41585-018-0003-3

Krausz C, Hoefsloot L, Simoni M, Tüttelmann F (2014) European academy of andrology; European molecular genetics quality network. EAA/EMQN best practice guidelines for molecular diagnosis of Y-chromosomal microdeletions: state-of-the-art 2013. Andrology 2(1):5–19. https://doi.org/10.1111/j.2047-2927.2013.00173.x

Kumar N, Singh AK (2015) Trends of male factor infertility, an important cause of infertility: a review of literature. J Hum Reprod Sci 8(4):191–196. https://doi.org/10.4103/0974-1208.170370

Lai TC, Roychoudhury S, Cho CL (2022) Oxidative stress and varicocele-associated male infertility. Adv Exp Med Biol 1358:205–235. https://doi.org/10.1007/978-3-030-89340-8_10

LaPointe S, Mehta A, Gaskins AJ (2023) Diet and recreational drug use in relation to male reproductive health. Fertil Steril 120(6):1089–1097. https://doi.org/10.1016/j.fertnstert.2023.10.013

Lazzarino G, Listorti I, Bilotta G, Capozzolo T, Amorini AM, Longo S, Caruso G, Lazzarino G, Tavazzi B, Bilotta P (2019) Water- and fat-soluble antioxidants in human seminal plasma and serum of fertile males. Antioxidants (Basel) 8(4):96. https://doi.org/10.3390/antiox8040096

Lee JD, Jeng SY, Lee TH (2006) Increased expression of hypoxia-inducible factor-1alpha in the internal spermatic vein of patients with varicocele. J Urol 175(3 Pt 1):1045–1048.; discussion 1048. https://doi.org/10.1016/S0022-5347(05)00417-9

Leisegang K, Sengupta P, Agarwal A, Henkel R (2021) Obesity and male infertility: mechanisms and management. Andrologia 53(1):e13617. https://doi.org/10.1111/and.13617

Liu J, Dai Y, Li R, Yuan J, Wang Q, Wang L (2023) Does air pollution exposure affect semen quality? Evidence from a systematic review and meta-analysis of 93,996 Chinese men. Front Public Health 11:1219340. https://doi.org/10.3389/fpubh.2023.1219340

Liu Q, Zhou WJ, Huang ZH, Huang XH, Wu J, Ji XR, Luo XF, Tang YL, Wang RJ, Fan LQ, Huang C, Zhu WB (2024) Exploring the impact of pericentric inversion of chromosome 9 on fertility in sperm donors. Asian J Androl 26(3):302–307. https://doi.org/10.4103/aja202369

Lo JO, Hedges JC, Girardi G (2022) Impact of cannabinoids on pregnancy, reproductive health, and offspring outcomes. Am J Obstet Gynecol 227(4):571–581. https://doi.org/10.1016/j.ajog.2022.05.056

Longini M, Moretti E, Signorini C, Noto D, Iacoponi F, Collodel G (2020) Relevance of seminal F_2-dihomo-IsoPs, F_2-IsoPs and F_4-NeuroPs in idiopathic infertility and varicocele. Prostaglandins Other Lipid Mediat 149:106448. https://doi.org/10.1016/j.prostaglandins.2020.106448
Mah PM, Wittert GA (2010) Obesity and testicular function. Mol Cell Endocrinol 316:180–186
Maione L, Dwyer AA, Francou B, Guiochon-Mantel A, Binart N, Bouligand J, Young J (2018) Genetics in endocrinology: genetic counseling for congenital hypogonadotropic hypogonadism and Kallmann syndrome: new challenges in the era of oligogenism and next-generation sequencing. Eur J Endocrinol 178(3):R55–R80. https://doi.org/10.1530/EJE-17-0749
Mannucci A, Argento FR, Fini E, Coccia ME, Taddei N, Becatti M, Fiorillo C (2022) The impact of oxidative stress in male infertility. Front Mol Biosci 8:799294. https://doi.org/10.3389/fmolb.2021.799294
Marinaro JA (2023) Sperm DNA fragmentation and its interaction with female factors. Fertil Steril 120(4):715–719. https://doi.org/10.1016/j.fertnstert.2023.06.001
Masenga SK, Kabwe LS, Chakulya M, Kirabo A (2023) Mechanisms of oxidative stress in metabolic syndrome. Int J Mol Sci 24(9):7898. https://doi.org/10.3390/ijms24097898
Mason MM, Clavijo RI (2023) Does varicocele treatment in adolescence improve fertility outcomes in adulthood? Eur Urol Focus 9(1):42–45. https://doi.org/10.1016/j.euf.2022.11.015
McElreavy K, Vilain E, Abbas N, Costa JM, Souleyreau N, Kucheria K, Boucekkine C, Thibaud E, Brauner R, Flamant F et al (1992) XY sex reversal associated with a deletion 5′ to the SRY "HMG box" in the testis-determining region. Proc Natl Acad Sci USA 89(22):11016–11020. https://doi.org/10.1073/pnas.89.22.11016
McSwiggin HM, O'Doherty AM (2018) Epigenetic reprogramming during spermatogenesis and male factor infertility. Reproduction 156(2):R9–R21. https://doi.org/10.1530/REP-18-0009
Micheli L, Cerretani D, Collodel G, Menchiari A, Moltoni L, Fiaschi AI, Moretti E (2016) Evaluation of enzymatic and non-enzymatic antioxidants in seminal plasma of men with genitourinary infections, varicocele and idiopathic infertility. Andrology 4(3):456–464. https://doi.org/10.1111/andr.12181
Milne GL (2017) Classifying oxidative stress by F_2-Isoprostane levels in human disease: the re-imagining of a biomarker. Redox Biol 12:897–898. https://doi.org/10.1016/j.redox.2017.04.028
Moretti E, Capitani S, Figura N, Pammolli A, Federico MG, Giannerini V, Collodel G (2009a) The presence of bacteria species in semen and sperm quality. J Assist Reprod Genet 26(1):47–56. https://doi.org/10.1007/s10815-008-9283-5
Moretti E, Pascarelli NA, Giannerini V, Geminiani M, Anichini C, Collodel G (2009b) 18, X, Y aneuploidies and transmission electron microscopy studies in spermatozoa from five carriers of different reciprocal translocations. Asian J Androl 11(3):325–332. https://doi.org/10.1038/aja.2008.31
Moretti E, Figura N, Campagna MS, Iacoponi F, Gonnelli S, Collodel G (2017) Infectious burden and semen parameters. Urology 100:90–96. https://doi.org/10.1016/j.urology.2016.10.032
Moretti E, Signorini C, Ferretti F, Noto D, Collodel G (2022) A study to validate the relevance of semen F_2-Isoprostanes on human male infertility. Int J Environ Res Public Health 19(3):1642. https://doi.org/10.3390/ijerph19031642
Moretti E, Signorini C, Corsaro R, Giamalidi M, Collodel G (2023) Human sperm as an in vitro model to assess the efficacy of antioxidant supplements during sperm handling: a narrative review. Antioxidants (Basel) 12(5):1098. https://doi.org/10.3390/antiox12051098
Muratori M, De Geyter C (2019) Chromatin condensation, fragmentation of DNA and differences in the epigenetic signature of infertile men. Best Pract Res Clin Endocrinol Metab 33(1):117–126. https://doi.org/10.1016/j.beem.2018.10.004
Muratori M, Marchiani S, Tamburrino L, Baldi E (2019) Sperm DNA fragmentation: mechanisms of origin. Adv Exp Med Biol 1166:75–85. https://doi.org/10.1007/978-3-030-21664-1_5
Neto FT, Bach PV, Najari BB, Li PS, Goldstein M (2016) Spermatogenesis in humans and its affecting factors. Semin Cell Dev Biol 59:10–26. https://doi.org/10.1016/j.semcdb.2016.04.009
Njagi P, Groot W, Arsenijevic J, Dyer S, Mburu G, Kiarie J (2023) Financial costs of assisted reproductive technology for patients in low- and middle-income countries: a systematic review. Hum Reprod Open 2023(2):hoad007. https://doi.org/10.1093/hropen/hoad007

Nowicka-Bauer K, Nixon B (2020) Molecular changes induced by oxidative stress that impair human sperm motility. Antioxidants (Basel) 9(2):134. https://doi.org/10.3390/antiox9020134

O'Flaherty C (2018) Peroxiredoxin 6: the protector of male fertility. Antioxidants (Basel) 7(12):173. https://doi.org/10.3390/antiox7120173

O'Flaherty C (2020) Reactive oxygen species and male fertility. Antioxidants (Basel) 9(4):287. https://doi.org/10.3390/antiox9040287

O'Flynn O'Brien KL, Varghese AC, Agarwal A (2010) The genetic causes of male factor infertility: a review. Fertil Steril 93(1):1–12. https://doi.org/10.1016/j.fertnstert.2009.10.045

Oghbaei H, Rastgar Rezaei Y, Nikanfar S, Zarezadeh R, Sadegi M, Latifi Z, Nouri M, Fattahi A, Ahmadi Y, Bleisinger N (2020) Effects of bacteria on male fertility: spermatogenesis and sperm function. Life Sci 256:117891. https://doi.org/10.1016/j.lfs.2020.117891

Olotu O, Ahmedani A, Kotaja N (2023) Small non-coding RNAs in male reproduction. Semin Reprod Med 41(6):213–225. https://doi.org/10.1055/s-0044-1779726

Paira DA, Silvera-Ruiz S, Tissera A, Molina RI, Olmedo JJ, Rivero VE, Motrich RD (2022) Interferon γ, IL-17, and IL-1β impair sperm motility and viability and induce sperm apoptosis. Cytokine 152:155834. https://doi.org/10.1016/j.cyto.2022.155834

Pecora G, Sciarra F, Gangitano E, Venneri MA (2023) How food choices impact on male fertility. Curr Nutr Rep 12(4):864–876. https://doi.org/10.1007/s13668-023-00503-x

Punab M, Poolamets O, Paju P, Vihljajev V, Pomm K, Ladva R, Korrovits P, Laan M (2017) Causes of male infertility: a 9-year prospective monocentre study on 1737 patients with reduced total sperm counts. Hum Reprod 32(1):18–31. https://doi.org/10.1093/humrep/dew284

Razi M, Malekinejad H (2015) Varicocele-induced infertility in animal models. Int J Fertil Steril 9(2):141–149. https://doi.org/10.22074/ijfs.2015.4234

Reynolds N, Cooke HJ (2005) Role of the DAZ genes in male fertility. Reprod Biomed Online 10:72–80

Ribeiro JC, Nogueira-Ferreira R, Amado F, Alves MG, Ferreira R, Oliveira PF (2022) Exploring the role of oxidative stress in sperm motility: a proteomic network approach. Antioxid Redox Signal 37(7–9):501–520. https://doi.org/10.1089/ars.2021.0241

Ritchie C, Ko EY (2021) Oxidative stress in the pathophysiology of male infertility. Andrologia 53(1):e13581. https://doi.org/10.1111/and.13581

Rodprasert W, Toppari J, Virtanen HE (2023) Environmental toxicants and male fertility. Best Pract Res Clin Obstet Gynaecol 86:102298. https://doi.org/10.1016/j.bpobgyn.2022.102298

Rogol AD (2023) Sex chromosome aneuploidies and fertility: 47,XXY, 47,XYY, 47,XXX and 45,X/47, XXX. Endocr Connect 12(9):e220440. https://doi.org/10.1530/EC-22-0440

Rozen SG, Marszalek JD, Irenze K, Skaletsky H, Brown LG, Oates RD, Silber SJ, Ardlie K, Page DC (2012) AZFc deletions and spermatogenic failure: a population-based survey of 20,000 Y chromosomes. Am J Hum Genet 91:890–896

Salas-Huetos A, Blanco J, Vidal F, Mercader JM, Garrido N, Anton E (2014) New insights into the expression profile and function of micro-ribonucleic acid in human spermatozoa. Fertil Steril 102(1):213–222.e4. https://doi.org/10.1016/j.fertnstert.2014.03.040

Salas-Huetos A, James ER, Aston KI, Jenkins TG, Carrell DT (2019) Diet and sperm quality: nutrients, foods and dietary patterns. Reprod Biol 19(3):219–224. https://doi.org/10.1016/j.repbio.2019.07.005

Salvio G, Ciarloni A, Cutini M, Delli Muti N, Finocchi F, Perrone M, Rossi S, Balercia G (2022) Metabolic syndrome and male fertility: beyond heart consequences of a complex cardiometabolic endocrinopathy. Int J Mol Sci 23:5497

Sansone A, Di Dato C, de Angelis C, Menafra D, Pozza C, Pivonello R, Isidori A, Gianfrilli D (2018) Smoke, alcohol and drug addiction and male fertility. Reprod Biol Endocrinol 16(1):3. https://doi.org/10.1186/s12958-018-0320-7

Santi D, Cignarelli A, Baldi M, Sansone A, Spaggiari G, Simoni M, Corona G (2024) The chronic alcohol consumption influences the gonadal axis in men: results from a meta-analysis. Andrology 12(4):768–780. https://doi.org/10.1111/andr.13526

Saxena R, de Vries JW, Repping S, Alagappan RK, Skaletsky H, Brown LG, Ma P, Chen E, Hoovers JM, Page DC (2000) Four DAZ genes in two clusters found in the AZFc region of the human Y chromosome. Genomics 67(3):256–267. https://doi.org/10.1006/geno.2000.6260
Schrott R, Murphy SK (2020) Cannabis use and the sperm epigenome: a budding concern? Environ Epigenet 6(1):dvaa002. https://doi.org/10.1093/eep/dvaa002
Sciorio R, Tramontano L, Adel M, Fleming S (2024) Decrease in sperm parameters in the 21st century: obesity, lifestyle, or environmental factors? An updated narrative review. J Pers Med 14(2):198. https://doi.org/10.3390/jpm14020198
Service CA, Puri D, Al Azzawi S, Hsieh TC, Patel DP (2023) The impact of obesity and metabolic health on male fertility: a systematic review. Fertil Steril 120(6):1098–1111. https://doi.org/10.1016/j.fertnstert.2023.10.017
Shi Q, Martin RH (2001) Aneuploidy in human spermatozoa: FISH analysis in men with constitutional chromosomal abnormalities, and in infertile men. Reproduction 121(5):655–666. https://doi.org/10.1530/rep.0.1210655
Sies H (2015) Oxidative stress: a concept in redox biology and medicine. Redox Biol 4:180–183. https://doi.org/10.1016/j.redox.2015.01.002
Sies H, Belousov VV, Chandel NS, Davies MJ, Jones DP, Mann GE, Murphy MP, Yamamoto M, Winterbourn C (2022) Defining roles of specific reactive oxygen species (ROS) in cell biology and physiology. Nat Rev Mol Cell Biol 23(7):499–515. https://doi.org/10.1038/s41580-022-00456-z
Signorini C, Moretti E, Collodel G (2020) Role of isoprostanes in human male infertility. Syst Biol Reprod Med 66(5):291–299. https://doi.org/10.1080/19396368.2020.1793032
Stokes VJ, Anderson RA, George JT (2015) How does obesity affect fertility in men—and what are the treatment options? Clin Endocrinol 82(5):633–638. https://doi.org/10.1111/cen.12591
Stouffs K, Lissens W, Verheyen G, Van Landuyt L, Goossens A, Tournaye H, Van Steirteghem A, Liebaers I (2004) Expression pattern of the Y-linked PRY gene suggests a function in apoptosis but not in spermatogenesis. Mol Hum Reprod 10:15–21
Szabó A, Váncsa S, Hegyi P, Váradi A, Forintos A, Filipov T, Ács J, Ács N, Szarvas T, Nyirády P, Kopa Z (2023) Lifestyle-, environmental-, and additional health factors associated with an increased sperm DNA fragmentation: a systematic review and meta-analysis. Reprod Biol Endocrinol 21(1):5. https://doi.org/10.1186/s12958-023-01054-0
Tian D, Yang C, Xie B, Li H, Li J, Yang D, Zhu Z (2023) Effects of varicocele surgical repair on serum hormone and inhibin B levels for patients with varicocele: a systematic review and meta-analysis. Am J Mens Health 17(5):15579883231199400. https://doi.org/10.1177/15579883231199400
Tiepolo L, Zuffardi O (1976) Localization of factors controlling spermatogenesis in the nonfluorescent portion of the human Y chromosome long arm. Hum Genet 34:119–124
Tomar A, Gomez-Velazquez M, Gerlini R, Comas-Armangué G, Makharadze L, Kolbe T, Boersma A, Dahlhoff M, Burgstaller JP, Lassi M, Darr J, Toppari J, Virtanen H, Kühnapfel A, Scholz M, Landgraf K, Kiess W, Vogel M, Gailus-Durner V, Fuchs H, Marschall S, Hrabě de Angelis M, Kotaja N, Körner A, Teperino R (2024) Epigenetic inheritance of diet-induced and sperm-borne mitochondrial RNAs. Nature 630(8017):720–727. https://doi.org/10.1038/s41586-024-07472-3
Tong J, Jiang J, Niu Y, Zhang T (2022) Do chromosomal inversion carriers really need preimplantation genetic testing? J Assist Reprod Genet 39(11):2573–2579. https://doi.org/10.1007/s10815-022-02654-2
Tunç E, Ilgaz S (2022) Robertsonian translocation (13;14) and its clinical manifestations: a literature review. Reprod Biomed Online 45(3):563–573. https://doi.org/10.1016/j.rbmo.2022.05.019
Ullah S, Ahmad S, Guo X, Ullah S, Ullah S, Nabi G, Wanghe K (2023) A review of the endocrine disrupting effects of micro and nano plastic and their associated chemicals in mammals. Front Endocrinol (Lausanne) 13:1084236. https://doi.org/10.3389/fendo.2022.1084236
Ustuner M, Yilmaz H, Yavuz U, Ciftci S, Saribacak A, Aynur BS, Yasar H, Culha MM (2015) Varicocele repair improves testicular histology in men with nonobstructive azoospermia. Biomed Res Int 2015:709452. https://doi.org/10.1155/2015/709452

Van Assche E, Bonduelle M, Tournaye H, Joris H, Verheyen G, Devroey P, Van Steirteghem A, Liebaers I (1996) Cytogenetics of infertile men. Hum Reprod 11(Suppl 4):1–24; discussion 25–6. https://doi.org/10.1093/humrep/11.suppl_4.1

Ventimiglia E, Capogrosso P, Boeri L, Serino A, Colicchia M, Ippolito S, Scano R, Papaleo E, Damiano R, Montorsi F, Salonia A (2015) Infertility as a proxy of general male health: results of a cross-sectional survey. Fertil Steril 104(1):48–55. https://doi.org/10.1016/j.fertnstert.2015.04.020

Verhaeghe F, Di Pizio P, Bichara C, Berby B, Rives A, Jumeau F, Sétif V, Sibert L, Rondanino C, Rives N (2020) Cannabis consumption might exert deleterious effects on sperm nuclear quality in infertile men. Reprod Biomed Online 40(2):270–280. https://doi.org/10.1016/j.rbmo.2019.11.002

Villaverde AISB, Netherton J, Baker MA (2019) From past to present: the link between reactive oxygen species in sperm and male infertility. Antioxidants (Basel) 8(12):616. https://doi.org/10.3390/antiox8120616

Vogt PH, Edelmann A, Kirsch S, Henegariu O, Hirschmann P, Kiesewetter F, Köhn FM, Schill WB, Farah S, Ramos C, Hartmann M, Hartschuh W, Meschede D, Behre HM, Castel A, Nieschlag E, Weidner W, Gröne HJ, Jung A, Engel W, Haidl G (1996) Human Y chromosome azoospermia factors (AZF) mapped to different subregions in Yq11. Hum Mol Genet 5(7):933–943. https://doi.org/10.1093/hmg/5.7.933

Walker WH (2022) Regulation of mammalian spermatogenesis by miRNAs. Semin Cell Dev Biol 121:24–31. https://doi.org/10.1016/j.semcdb.2021.05.009

Wang S, Zhang K, Yao Y, Li J, Deng S (2021) Bacterial infections affect male fertility: a focus on the oxidative stress-autophagy Axis. Front Cell Dev Biol 9:727812. https://doi.org/10.3389/fcell.2021.727812

Wang X, Ramat A, Simonelig M, Liu MF (2023) Emerging roles and functional mechanisms of PIWI-interacting RNAs. Nat Rev Mol Cell Biol 24(2):123–141. https://doi.org/10.1038/s41580-022-00528-0

Wanjari UR, Gopalakrishnan AV (2023) A review on immunological aspects in male reproduction: an immune cells and cytokines. J Reprod Immunol 158:103984. https://doi.org/10.1016/j.jri.2023.103984

Wood GJA, Cardoso JPG, Paluello DV, Nunes TF, Cocuzza M (2021) Varicocele-associated infertility and the role of oxidative stress on sperm DNA fragmentation. Front Reprod Health 3:695992

World Health Organization (2021) WHO laboratory manual for the examination and processing of human semen, 6th edn

Yan Y, Yang X, Liu Y, Shen Y, Tu W, Dong Q, Yang D, Ma Y, Yang Y (2017) Copy number variation of functional RBMY1 is associated with sperm motility: an azoospermia factor-linked candidate for asthenozoospermia. Hum Reprod 32:1521–1531

Zhang X, San Gabriel M, Zini A (2006) Sperm nuclear histone to protamine ratio in fertile and infertile men: evidence of heterogeneous subpopulations of spermatozoa in the ejaculate. J Androl 27(3):414–420. https://doi.org/10.2164/jandrol.05171

Zhao L, Pang A (2020) Effects of metabolic syndrome on semen quality and circulating sex hormones: a systematic review and metaanalysis. Front Endocrinol (Lausanne) 11:544553

Zini A, Defreitas G, Freeman M, Hechter S, Jarvi K (2000) Varicocele is associatedwith abnormal retention of cytoplasmic droplets by human spermatozoa. Fertil Steril 74:461–464

Zylbersztejn DS, Andreoni C, Del Giudice PT, Spaine DM, Borsari L, Souza GH, Bertolla RP, Fraietta R (2013) Proteomic analysis of seminal plasma in adolescents with and without varicocele. Fertil Steril 99:92–98

Non-systematic and Systematic Defects of Human Spermatozoon

5

Contents

Core Messages

The normal spermatozoon is a differentiated cell with a head and a tail. The head comprises a cap-like acrosome containing enzymes to penetrate the membranes of the oocyte and a well-shaped nucleus. For the spermatozoon to be able to move and fertilize, the tail must be well organized in all its structures, the axoneme, the accessory fibers, the fibrous sheath, and the mitochondria. All sperm components can be altered and therefore functionality compromised. This chapter discusses sperm defects, including their classification into systematic and non-systematic alterations, their potential causes, and the current state of knowledge in this field. New acquisitions are fundamental for increasing therapies and positive outcome in assisted reproduction techniques.

Abbreviations

ART	Assisted reproductive technologies
AURKC	Aurora Kinase C

E. Moretti, G. Collodel, *The Human Spermatozoon*,
https://doi.org/10.1007/978-3-031-97868-5_5

DFS	Dysplasia of the fibrous sheath
FRAC	Fluorescence-based ratiometric assessment of centrioles
ICSI	Intracytoplasmic sperm injection
IDA	Inner dynein arm
LM	Light microscopy
MMAF	Multiple morphological abnormalities of the sperm flagella
mtDNA	Mitochondrial DNA
NFSA	Non-specific flagellar anomalies
ODA	Outer dynein arm
PCD	Primary ciliary dyskinesia
ROS	Reactive oxygen species
SEM	Scanning electron microscopy
SEPT	Septin
TEM	Transmission electron microscopy

5.1 Anatomy of Human Sperm Defects

David Mortimer, in his article "The functional anatomy of the human spermatozoon: relating ultrastructure and function" (Mortimer 2018), makes the following key observation: "*The Internet, magazine articles, and even biomedical journal articles are full of cartoons of spermatozoa that bear minimal resemblance to real spermatozoa, especially human spermatozoa, and this had led to many misconceptions about what spermatozoa look like and how they are constituted. In the author's experience a surprising number of biomedical reproductive biologists, especially those working in assisted reproductive technologies (ART), have limited knowledge of sperm functional anatomy and how it underpins and influences sperm function.*"

In few words, human spermatozoa are routinely used in the field of ART without knowing in depth how they are structured, and the real morphological defects they carry.

In laboratory practice, light microscopy (LM) is the most widely used technique to evaluate sperm morphology in semen samples smeared on the glass slides and stained. LM enables a very simple analysis of sperm morphology but has technical limitations because of its low resolving power and inability to highlight numerous morphological defects of sperm structures and organelles.

Transmission electron microscopy (TEM) represents one of the best tools, allowing for the examination of the cell ultrastructure at high magnification, identifying the different structures in normal and altered forms. Thus, TEM is important for the assessment of sperm defects that are responsible for reduced fertilizing potential. Ultrastructural studies integrated with molecular techniques enable a detailed characterization of sperm defects from structural, molecular, and functional points of view (Chemes and Alvarez Sedo 2012; Chemes and Rawe 2003; Moretti et al. 2016, 2022).

This methodological approach led to the establishment, approximately three decades ago, of the concept of sperm pathology as a distinct discipline dedicated to the study of structural and functional alterations in spermatozoa. The introduction of sperm pathology originated the classification of ultrastructural alterations in "non-systematic or non-monomorphic" and "systematic or monomorphic" sperm defects (Baccetti et al. 2001; Chemes and Rawe 2003).

Non-monomorphic sperm defects include a heterogeneous combination of different alterations, potentially affecting all the sperm structures. Since these anomalies can be caused by several pathologies (varicocele, hormonal imbalance, infections, chronic diseases), environmental factors and lifestyle, they can potentially be treated. These defects have a variable prevalence in the different patients and in the same subject through his clinical evolution, and are present in the fertile male population in a low percentage.

Systematic or monomorphic sperm defects are characterized by a principal anomaly present in most spermatozoa of all patients suffering the same disorder (Baccetti et al. 2001; Chemes and Rawe 2003) and when they were discovered, they were immediately considered to have a genetic origin, even though the genes were unknown. These defects are more frequent in subjects with a history of consanguinity and show family clustering. Head, head-neck attachment, and flagellum are all regions of the spermatozoon that can be affected by these monomorphic sperm defects.

Over the years, the scanning electron microscopy (SEM) was a relevant support of TEM in the characterization of systematic sperm defects (Chemes and Rawe 2003).

This classification, based on morphological characteristics, is still currently valid; however, it is blended with the recent genetic studies that are opening new perspectives in the study of sperm defects. In fact, significant advances have been made in the field of the genetics of sperm defect. Clinical studies that correlate some infertility case with the loss of function of one or more genes are now present in the literature. For these kinds of studies, the correct selection of patients and a deep study of their anamnesis are relevant because the concomitant presence of other pathologies and conditions can distort the results.

In addition, a growing number of genes that are necessary for reproduction are being identified via gene-knockout/mutation studies in mice, opening up new possibilities for investigating the function of these in male infertility field (Jiao et al. 2021).

For these reasons, this chapter summarizes the ultrastructural characteristics of sperm defects, integrating them with the modern genetic studies that greatly facilitated the identification of infertility-associated gene mutations in patients over the past 20 years.

5.2 Alterations of Sperm Head

The differences in head dimension and structure ("amorphous," "pyriform," and "tapered" heads) are the basis for morphological classification (WHO 2021) of the defects but are not associated with a specific sperm anomaly (Chemes and Alvarez Sedo 2012).

TEM analysis is important in the evaluation of sperm nuclear shape and in the identification of binucleated and multinucleated spermatozoa, an alteration particularly observed in individuals with high serum levels of prolactin (Baccetti et al. 1978) and in cases of chromosomal translocation (Baccetti et al. 2005a; Moretti et al. 2009).

The most relevant alterations affecting sperm heads involve the chromatin and the acrosome, both sperm component that can be analyzed by TEM.

In testicular germ cells up to early round spermatids, the DNA molecule is associated with histones, as in the somatic cells. During spermiogenesis, the histones are replaced by transition proteins and then by protamines (see Sect. 2.4.3): DNA-protamine fibers are extremely compacted assuring stability and protection to the male genome during the long way to oocyte (Ribas-Maynou et al. 2022).

Alterations in chromatin condensation are one of the most frequent findings in teratozoospermia and often associated with irregular nuclear profiles (Chemes 2018). In particular, the immature chromatin has a granular texture, and it is frequently a characteristic of elliptical or spherical sperm heads (Collodel et al. 2013).

The chromatin of necrotic spermatozoa is characterized by a peculiar, disrupted texture, and it is often associated with broken plasma and acrosomal membranes, mitochondrial alterations and degeneration of axonemal and periaxonemal structures (Chemes and Rawe 2003).

In any case, mature sperm with condensed chromatin and sperm with immature or necrotic chromatin can present a sort of series of holes, first called "vacuoles." These vacuoles can be observed in the head of sperm from fertile men and infertile patients and vary in number, position, and size (Gómez-Torres et al. 2023). By TEM analysis, these "vacuoles" appear as clear granular or filamentous spaces within the chromatin of the sperm nucleus not surrounded by a plasma membrane; for this reason, the term "vacuoles" was reputed incorrect. Chemes and Alvarez Sedo (2012) suggested the term "lacunar defects" or "lacunae," and recently Gómez-Torres et al. (2023) demonstrated that these human sperm head vacuoles originate from nuclear invaginations and contain perinuclear theca components; for these reasons, they proposed the new definition "nuclear invaginations." Other authors suggested that this sperm dysfunction is related to sperm DNA damages caused by abnormal chromatin packaging (Boitrelle et al. 2011; Franco Jr et al. 2012; Oliveira et al. 2010), aneuploidy (Perdrix et al. 2011), and instability of the genome (Berkovitz et al. 2006). Also Pastuszek et al. (2017) found a direct correlation between medium- and large-sized nuclear "vacuoles" and high levels of DNA fragmentation. Moreover, sperm vacuoles seem to be related to male fertility potential because they have been associated with implantation failures or miscarriages, even though different studies have also argued that sperm nuclear vacuoles result from a natural physiological process (Gómez-Torres et al. 2023; Tanaka et al. 2012).

The acrosome (see Sect. 3.2.1) derives from vesicles produced by the Golgi apparatus of early spermatids; these vesicles fuse forming a large pro-acrosomal vesicle, which migrates, attaches to the nuclear membrane, and covers 75% of the sperm head (Chemes 2018). As the acrosome grows, the nucleus is modeled by the forces exerted by the microtubular manchette (see Sect. 2.4.3). These two structures are related since acrosomal alterations are frequently concomitant with nuclear defects and the lack and the insufficient development of the acrosome can severely compromise the fertilizing ability (Moretti et al. 2016).

This relationship between the development of the acrosome, the nucleus, and the manchette is demonstrated by a pathological condition called globozoospermia.

A peculiar defect was described in the past: the crater defect observed in several domestic species and described also in human spermatozoa by Baccetti et al. (1989). The defect occurs during spermiogenesis and typically appears as a single deep invagination into the acrosome and nucleus in all the spermatozoa of the affected individual.

5.2.1 Globozoospermia

Globozoospermia is a rare genetic defect (systematic defect) characterized by round-headed spermatozoa without acrosome, nucleus with immature chromatin (Yassine et al. 2015), absence of a post-acrosomal sheath (Fig. 5.1a, b) and increased susceptibility to non-enzymatic fatty acid oxidation (Moretti et al. 2019). Recent observations (Moretti et al. 2019; Ricci et al. 2015) described new morphological alterations associated with globozoospermia in human spermatozoa such as coiled tail around the head (Fig. 5.1a) that represents a crucial step of defective sperm maturation.

A similar phenotype was observed in male mice whose gene encoding Golgi-associated PDZ and coiled-coil motif containing protein was deleted (Yao et al. 2002).

The classic form is characterized by 100% round-headed spermatozoa (total globozoospermia), which are unable to penetrate the oocyte and thus cause primary infertility. "Partial globozoospermia" is referred to a mosaic of normal and round-headed sperm, considering that also the semen of general fertile population can present a proportion up to 9% of spermatozoa with globozoospermic characteristics (Auger et al. 2016).

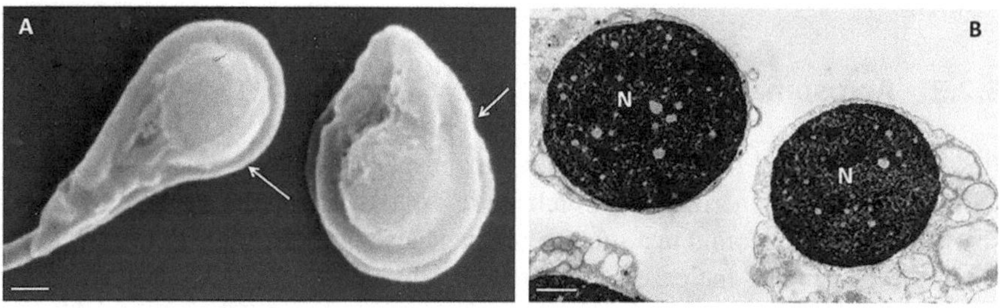

Fig. 5.1 Scanning (**A**) and transmission (**B**) electron microscopy micrographs of spermatozoa with globozoospermia. In (**A**) the spermatozoa show evident round heads with coiled tails (arrows); in (**B**) the spermatozoa show round nuclei (**N**) with uncondensed chromatin. It is noteworthy the total lack of acrosome. Bars: 1 μm

Several familial cases of globozoospermia suggested a genetic contribution to this disorder. This hypothesis was supported by the description in the literature of knock out mice with a globozoospermia-like phenotype lacking different genes. Numerous proteins are involved in the different steps of acrosome formation and their absence leads to defective acrosome biogenesis and globozoospermia or globozoospermia-like phenotypes (Beurois et al. 2020; Coutton et al. 2015). The alterations of these proteins during acrosomal biogenesis interfere with the perinuclear localization and the shaping of the acrosomal vesicle, not with its formation. Therefore, the acrosome/acrosomal vesicles are removed in the residual body when spermatozoa are released from seminiferous epithelium (Chemes and Alvarez Sedo 2012).

The main genes involved in the pathogenesis of globozoospermia are *DPY19L2*, *SPATA16*, *PICK1*, *GGN*, *SPACA1*, *ZPBP*, *CCDC62*, and *CCNB3*. This panel of genes should be evaluated in patients affected by globozoospermia (Crafa et al. 2023). In particular, next generation sequencing allows identification of several mutations of the *DPY19L2* gene, the main gene associated with the globozoospermia phenotype (Fesahat et al. 2020). In cases of globozoospermia, the impossibility of a natural fertilization is evident and intracytoplasmic sperm injection (ICSI) represents the only chance to achieve pregnancy. However, spermatozoa from globozoospermic patients suffer fromdepletion of the oocyte activator phospholipase Cζ and post-acrosomal sheath WW binding-domain protein, located in the inner acrosomal membrane and in perinuclear theca, and this causes several problems in ICSI outcome. Several strategies to overcome this problem and improve the outcome of assisted fertilization in these patients have been proposed. The most common ones include ICSI with artificial oocyte activation (with Ca^{2+} ionophores or strontium), a way to artificially induce calcium release to trigger oocyte activation and intracytoplasmic morphologically selected sperm injection, which allows a detailed morphological sperm examination and the identification of acrosomal remnants carrying cytosolic factors essential for oocyte activation (Crafa et al. 2023).

In addition, spermatozoa of patients with globozoospermia show an increased rate of DNA fragmentations and aneuploidies that should be evaluated for a correct management of these infertile patients (De Braekeleer et al. 2015).

5.2.2 Acrosomal Hypoplasia

Broad ultrastructural investigations in teratozoospermia have revealed spermatozoa with small, detached acrosomes, which are referred to acrosomal hypoplasia. These spermatozoa can be found in a small percentage also in normal ejaculates; however, when the percentage increases and acrosomal hypoplasia is associated to round/elliptical nuclei, problem in oocyte activation can be present (Chemes and Alvarez Sedo 2012; Moretti et al. 2016).

In 1991, Baccetti and colleagues described a defect occurring in the total sperm population of two brothers. These spermatozoa showed elliptical nucleus devoid of post-acrosomal sheath. The acrosome was small in size and often detached from the nucleus; the defect was named "miniacrosome" (Baccetti et al. 1991).

Recently, it was observed that mutations of tektins 3 (TEKT3), a protein expressed in the axoneme, reduce sperm motility, and contribute to severe acrosomal hypoplasia in human spermatozoa, compromising their normal function (Liu et al. 2023a). Interestingly, Wang et al. (2024) observed in KO mice for coiled-coil domain-containing 38 (Ccdc38), spermatozoa with acrosomal hypoplasia and typical multiple morphological abnormalities of the sperm flagella (MMAF) phenotypes. Ccdc38 knockout causes a decreased level of TEKT3 in testes and an aberrant distribution of TEKT3 in sperm.

5.2.3 Macrozoospermia

Macrozoospermia or large-headed multiflagellar spermatozoa is a rare sperm anomaly characterized by the presence of near 100% of large-headed spermatozoa with multiple irregular nuclei, and several flagella, and it is one of the most severe phenotypes of male infertility. Meiotic segregation studies revealed that over 90% of spermatozoa were aneuploid, mainly diploid (De Braekeleer et al. 2015; Ray et al. 2017).

The description of familial cases with consanguineous parents suggested a genetic cause with an autosomal recessive inheritance. The first and main gene discovered was Aurora kinase C (*AURKC*), whose mutations were found in patients with macrozoospermic spermatozoa. AURKC is a member of the family (with AURKA and AURKB) of serine/threonine kinases known as Aurora kinases. AURKC is highly expressed in the testis, especially in spermatocytes that are dividing. The *AURKC* gene is the ideal candidate for this defect because it is expressed preferentially in male germ cells and is involved in chromosomal segregation and cytokinesis (Fellmeth et al. 2016). These observations are congruent with the abnormal sperm morphology and cytogenetic content of large-headed spermatozoa (Beurois et al. 2020; Coutton et al. 2015). ICSI should be contraindicated for all mutated patients because the produced gametes are always aneuploid.

The identification of *AURKC* and other genes as *ZMYND15* and *NUP210L* have been recently found to cause macrozoospermia; for this reason, Coudert et al. (2023) proposed an optimized strategy for the routine genetic investigation of macrozoospermic patients according to the percentage of large-headed spermatozoa.

5.3 Defects of Head-Tail Connecting Region

The head-tail connecting region or neck (for details, see Sect. 3.3) contains the centrosome that plays an important role in the fertilization process, because human oocyte lacks the centrioles necessary for the arrangement of mitotic spindle and the embryo mitotic division.

In recent decades, this field of research has proven particularly dynamic in the pursuit of genes responsible for sperm defects. One of the first alterations mentioned in the literature was the so-called pin-headed spermatozoa described as spermatozoa with an extremely small head. Then, it was demonstrated that these spermatozoa were devoid of head, they were "decapitated," "acephalic" spermatozoa and what was identified as "small head" was a cytoplasmic droplet (Chemes 2018).

The defects of the connecting piece essentially derive from an incorrect head-tail attachment. During spermiogenesis, the centrosome physiologically reaches the base of the nucleus to develop the flagellum (see Sect. 2.4.3). If this structure attaches the nucleus in the lateral position, the correct alignment between head and tail is lost, and the head results bent at different angles with respect to the tail; this particular configuration makes the head-tail junction very fragile.

The extreme manifestation of this phenomenon brings about the complete detachment of the head from the tail: "head-tail detachment"/"acephalic sperm."

Since this defect is highly stable over time, unresponsive to treatment, and exhibits similar characteristics across affected individuals, it can be classified as a systematic sperm defect with a likely genetic basis. In the last decades, many efforts have been made to study the numerous proteins (Avidor-Reiss et al. 2020; Fishman et al. 2018; Galletta et al. 2020), that, together with the centrioles, link the head to the tail, and that, if mutated, can be responsible for sperm head-tail detachment. Several mutated genes, studied in animal model and humans (Wang et al. 2022a), cause this particular phenotype; the purpose of this section is to focus on target genes having an impact on ICSI outcome.

ICSI is the only technique by which immotile, abnormal spermatozoa can fertilize the oocyte, but in presence of sperm head-tail detachment is extremely complex.

Following the new "centriole remodeling hypothesis" (see Sects. 2.4.3 and 3.3; Fishman et al. 2018; Avidor-Reiss et al. 2020), when the tailless head is injected into the oocyte at least one of the two centrioles, proximal centriole and modified distal centriole, must be present to arrange the mitotic spindle of the zygote. Recently, Nie et al. (2020), based on TEM observations, classified into different subgroups the sites of break of connecting piece in acephalic sperm and related them to mutated genes (Fig. 5.2). In the subgroup 1, the site of break occurs between the proximal centriole located in the head and the modified distal centriole remaining in the tail; no mutations have been identified yet. In the subgroup 2, the head retains the two centrioles and the identified involved genes are *HOOK1, SUN5* (Elkhatib et al. 2017; Zhu et al. 2016), and *PMFBP1* (Liu et al. 2021; Sha et al. 2019). In the subgroup 3, the tail contains both proximal and atypical distal centrioles; *TSGA10* (Toyama et al. 2000) and *BRDT* (Li et al. 2017) are the involved genes.

This classification system is meaningful and has implications on the ICSI outcome. Cases included in subgroup 2 achieved the best ICSI results because the injected head contains both centrioles, the proximal centriole and the atypical distal centriole, and those included in subgroup 3 are at real risk to fail clinical pregnancy because the centrioles are both located in the tail.

Several studies are focused on the discovery of new genes associated with the acephalic sperm defects (Avidor-Reiss et al. 2020; Sha et al. 2020). For this reason,

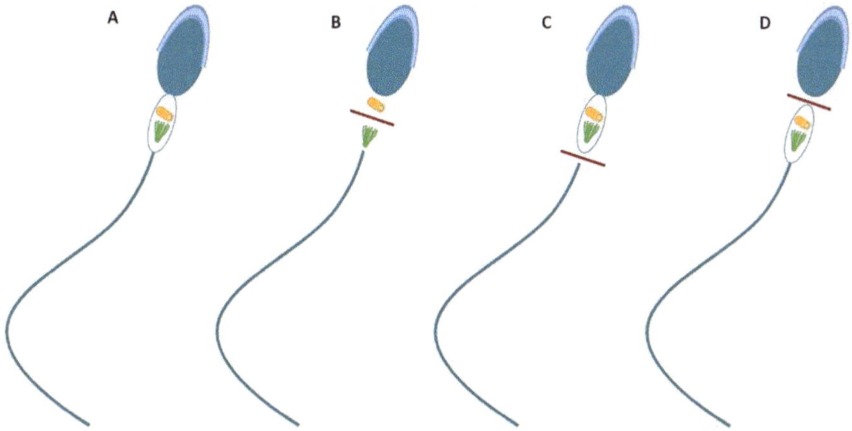

Fig. 5.2 Simplified representation of human spermatozoa and the possible head-tail points of breakage. Following the "centriole remodeling hypothesis," the focus is on the proximal centriole (yellow) and the atypical distal centriole (green). The striated columns and the capitulum, which are part of the connecting piece, are omitted. A normal sperm (**A**) shows the connecting piece linking the head and the tail. In (**B**), the head and the tail are broken, and the head contains the proximal centriole, the tail the atypical distal centriole (subgroup 1). In (**C**) both centrioles are included in the head separated from the tail (subgroup 2). In (**D**) both centrioles are included in the broken tail (subgroup 3)

the screening of these known and new mutations in infertile men attending assisted fertilization can clarify the genetic cause of the sperm defect, allowing a better management and the counseling of the couple (Mazaheri Moghaddam et al. 2021). The problems in the connecting piece area are not only related to the head-tail detachment that is an evident alteration. In the clinical practice, the investigation of the connecting piece of human sperm is technically difficult, because the best way to study this area is the use of TEM. Recently, fluorescence-based ratiometric assessment of centrioles (FRAC, Turner et al. 2021) was developed as a potential method to fill this gap. FRAC is based on immunofluorescence, image quantification and ratio calculation of the different fluorescent centriolar biomarkers. By this method, Jaiswal et al. (2022) demonstrated that some cases of unexplained infertility are due to centriolar alterations, underlining the urgent need to find methods to study this region.

An understudied structure present in the connecting piece is the microtubular centriolar adjunct that originates from the proximal centriole during spermiogenesis and is reduced in mature human spermatozoa (see Sect. 2.4.3). The persistence of centriolar adjunct in human spermatozoa can indicate incomplete maturation of the sperm connecting piece. The data on this topic is very limited. Garanina et al. (2019) examined, at TEM levels, human spermatozoa of fertile and infertile subjects and found that the centriolar adjuncts were increased in length in the spermatozoa of the two patients with unexplained infertility. Soon after, Moretti et al. (2021) described a case of a patient who obtained aneuploidy embryos naturally and with ART. The spermatozoa of this subject showed a combination of elongated and immature centriolar adjunct, altered sperm morphology, increased rate of aneuploidy.

These few data suggest that immature and altered connecting piece negatively influences zygote formation and development. Therefore, centriolar adjunct length can be potentially used as a complementary criterion to establish the degree of immaturity in human spermatozoa.

5.4 Alterations of Midpiece and Mitochondria

The mitochondria are structured as a helix around the axoneme during spermiogenesis. The role of mitochondria in spermatozoa is evident: they essentially are the central player in providing energy, producing ATP via oxidative phosphorylation, for the flagellar motility. Moreover, in the spermatozoon, mitochondria regulate redox equilibrium, calcium homeostasis, apoptotic pathways, processes needed for capacitation, acrosome reaction, and sperm–oocyte fusion (Tesarik and Mendoza-Tesarik 2023). Mitochondria are reactive oxygen species (ROS) producers due to the electron transport chain located in the inner mitochondrial membrane (see Sect. 3.3). The spermatozoon needs ROS that act as signal molecules for motility, capacitation, acrosomal reaction, and fertilization; however, a ROS overproduction and/or a dysregulation of the antioxidant scavenging system, becomes harmful in cells. At this purpose, mitochondrial dysfunctions cause the redox balance alterations inducing oxidative stress that causes aging and metabolic alterations that are prevented and controlled by a family of proteins called sirtuins (Di Emidio et al. 2021).

An ultrastructural study carried out in different mammalian sperm by Gu et al. (2019) have shown that the structure of midpiece influences sperm motility and metabolism. In particular, the study revealed positive associations among mitochondria, outer dense fibers and the length of the flagellum, and the sperm velocity, suggesting a close relationship between morphology and physiology.

Results on structural and ultrastructural mitochondrial sperm defects are not frequent in the literature. Essentially, they regard the length of the midpiece, which can be more extended than normal and associated with other flagellar anomalies (Moretti et al. 2008; Rawe et al. 2007). In contrast, Rawe et al. (2007) reported a case characterized by absent or markedly shortened midpieces, attributed to the absence or extreme scarcity of mitochondria. In case of lack of mitochondria, the midpiece appeared thin, weak, and susceptible to acute bending. Also, Mundy et al. (1995) reported short midpieces characterized by reduced mitochondrial helix gyres. In addition, abnormal assembly of the mitochondrial helix and structural defects of mitochondria that appear swollen with few cristae are sperm defects that could be linked to the fertilization rates (Baccetti et al. 2002).

Mitochondria depletion was observed also in the systematic sperm defect "dysplasia of fibrous sheath" (DFS, see Sect. 5.5.3). DFS spermatozoa show a short and thick tail with hyperplastic fibrous sheaths and few mitochondria near the nucleus (Chemes 2018).

In any case, spermatozoa with severe midpiece defects are immotile or show a severely reduced motility. An altered functioning of mitochondria has been found in spermatozoa of infertile patients; altered electron transport chain, oxygen consumption and membrane mitochondrial potential that are correlated with motility, vitality, capacitation (Ferramosca et al. 2012) and with acrosome reaction ability in humans (Gallon et al. 2006). In patients with asthenozoospermia, modifications of several mitochondrial proteins involved in the metabolism and energy production in spermatozoa were described and extensively revised by Boguenet et al. (2021). Therefore, reduced motility can be caused by dysfunctional mitochondria and insufficient production of energy. As an example, mitofusin-2, a protein involved in mitochondrial fusion phenomena and in the preservation of mitochondrial morphology and function, is expressed at reduced level in the sperm midpiece of asthenozoospermic patients (Fang et al. 2018).

Mitochondria are also a protagonist in regulating the intrinsic pathway of apoptosis. Although the existence of apoptosis in spermatozoa is a controversial issue, essentially due to the particular architecture of these cells, various studies have reported the typical apoptotic marker as membrane phosphatidylserine externalization, loss of mitochondrial integrity, caspase activation, and DNA damage (Boguenet et al. 2021). These apoptotic markers have been found in spermatozoa from infertile men and immature sperm, whereas were absent in mature sperm from fertile men (Grunewald et al. 2017).

Mitochondrial DNA (mtDNA) represents another aspect associated with sperm physiology and quality. MtDNA is more prone to be damaged by ROS than the nuclear genome, point mutations, and rearrangement correlated with sperm dysfunctions and infertility (Chianese and Pierantoni 2021). A low mtDNA copy number has been suggested as an indicator of good-quality sperm because in spermatozoa of infertile patients affected by asthenozoospermia and oligozoospermia, an increase in mtDNA copy number was described and associated with impaired fertilization and decreased probability of pregnancy (Rosati et al. 2020).

As already mentioned, the transition between midpiece and principal piece is defined by the annulus, a septin (SEPT)-based protein structure (see Sect. 3.3). Septin mutations cause defects on midpiece. For example, SEPT4 variants cause the presence of thin midpiece associated with asthenoteratozoospermia (Lhuillier et al. 2009; Wang et al. 2022b).

5.5 Alterations of Sperm Tail: Focus on Axonemal and Periaxonemal Structures

Morphological and/or functional defects of the sperm flagellum are responsible for asthenozoospermia, one of the main causes of male infertility. In particular, in this section, defects of the axonemal and periaxonemal structures will be illustrated, trying to define the more frequent heterogeneous acquired anomalies (defined as non-systematic sperm defects) and the homogeneous irreversible defects affecting most spermatozoa (defined as systematic sperm defects). For this classification, the ultrastructural analysis by TEM played a relevant role (Chemes 2018). However, TEM is not generally available in the laboratories where routine semen analysis is performed, and sperm morphology is evaluated by LM, which is a technique with evident limits in the study of axonemal and periaxonemal structures. Some attempts to correlate morphological characteristics assessed by LM and axonemal defects analyzed by TEM were performed (Mitchell et al. 2015), but ultrastructural studies are essential in many cases of asthenozoospermia. The morphological ultrastructural characterization of non-systematic and systematic sperm tail defects takes several decades, and it is now almost complete. In the last years, the interest was shifted to finding of the potential genes whose mutations cause systematic defects.

In this section, the morphological defects are presented in a didactic manner, accompanied by a general overview of recently identified genes associated with these abnormalities.

5.5.1 Non-specific Flagellar Anomalies (NSFA)

A regular structure of sperm axonemal and periaxonemal components is necessary for the normal and efficient motility needed to reach and fertilize the oocyte. Many sperm with non-specific flagellar anomalies (NFSA) have been observed in the ejaculates of the majority of patients with severe asthenozoospermia, and, in a consistently lower percentage, also in fertile men (Chemes 2000).

For this reason, the study of these anomalies in each patient could be relevant for making a reliable diagnosis, although difficult for the extreme specialization of the ultrastructural analysis (Chemes and Rawe 2003). NSFA affect axonemal or periaxonemal structures in different percentages of ejaculated spermatozoa without a definite pattern. The incidence of NSFA waves during the clinical evolution of the patient and it differs among patients.

Defects affecting microtubules and other axonemal proteins do not change the diameter of the flagellum, which seems normal and well-structured at LM level, and the abnormalities are not assessable on semen smears. Therefore, TEM analysis is required to observe and quantify the sperm tail structural defects leading to different degrees of asthenozoospermia.

Axonemal anomalies mainly consist of alterations of the number of microtubular doublets and their localization different from the normal 9 + 2 organization (Chemes 2018; Chemes and Rawe 2003; Wilton et al. 1992).

Abnormalities of the periaxonemal elements as the number, structure, and organization of outer dense fibers and the spatial organization of fibrous sheath lead to motility disturbances and flagellar diskinesia (Escalier 2003). The kind of defects are secondary to different andrological conditions usually affecting sperm motility such as varicocele, cryptorchidism, genitourinary infections, or idiopathic infertility. Examining the ejaculates of subjects with asthenozoospermia, a frequent finding is the presence of a variable percentage of coiled tails often embedded in large cytoplasmic residues with axonemal and periaxonemal structures severely compromised. The high presence of cytoplasmic residues is frequent in immature spermatozoa observed in case of varicocele and chromosomal alterations (Baccetti et al. 2002).

Extremely severe asthenozoospermia or total sperm immotility are disorders that, rarely, can be caused also by the presence of systematic sperm defects, when a homogenous phenotype is present in the vast majority of a sperm population in infertile patients.

5.5.2 Primary Ciliary Diskinesia

Primary ciliary dyskinesia (PCD) is a multisystemic disorder caused by motility defects of cilia and flagella, structures that have in common the axoneme, even though recent studies in humans and in animal models have shown differences in specific component within cilia and sperm tail axoneme (Sironen et al. 2020). These differences could represent the reason not all men with PCD are infertile and, on the other hand, axonemal defects can be present only in the sperm flagella (Jiao et al. 2021).

Two reports from Scandinavian researchers in 1975 revealed the alteration of the structure and function of cilia and flagella in patients affected by chronic respiratory disease, sperm immotility and *situs viscerum inversus* (Afzelius et al. 1975; Pedersen and Rebbe 1975). These findings introduced the concept of the immotile cilia syndrome, now named as PCD, where "primary" indicates the genetic nature of the syndrome and dyskinesia the concept of defective motility (Chemes 2018). It is now known that about half of PCD carriers have Kartagener syndrome characterized by a mirror image orientation of the heart and other internal organs. PCD usually shows an autosomal recessive inheritance, but rare cases of autosomal dominant and X-linked inheritance have been observed. Genetic counseling should be mandatory for couples pursuing ART who are affected by PCD because they are at increased risk of transmitting defects such as abnormal position of organs and chronic respiratory tract infections (Krausz and Riera-Escamilla 2018).

During semen analysis at LM level, PCD is characterized by spermatozoa with normal flagellar length and severe asthenozoospermia or total immotility associated with a broad spectrum of situations ranging from normal sperm concentration to oligozoospermia, and spermatozoa with an apparently normal morphology and teratozoospermia (Cannarella et al. 2020; Sironen et al. 2020). TEM analysis is important to confirm the ultrastructural defects in these spermatozoa characterized by axonemes lacking the outer (ODA), inner (IDA), or both dynein arms (Fig. 5.3), by the absence of one or two central microtubules or radial spokes and transposed microtubules (Chemes et al. 1990; Escalier and David 1984). ODA and IDA, linked to microtubule A of the doublet and projecting toward microtubule B of adjacent doublet, produce the force for motility (Sironen et al. 2020).

PCD is then caused by genetic variants in genes coding for proteins involved in motile cilia and flagella structure, formation, and function. Mutations in over 40 genes have been identified so far; however, the exact effect of these mutations on

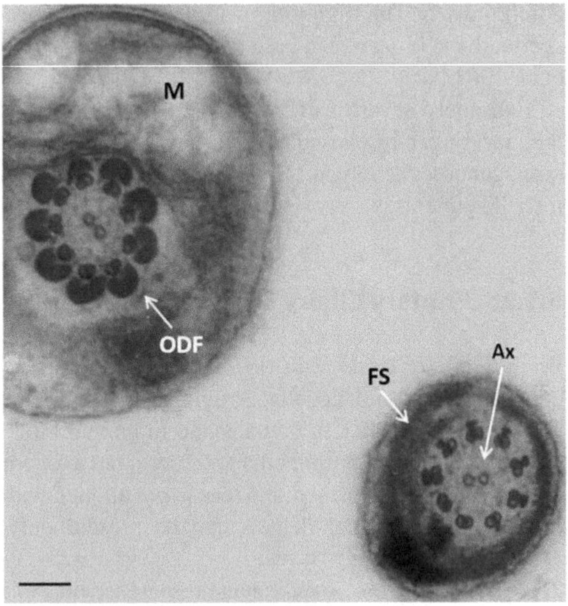

Fig. 5.3 Transmission electron microscopy micrograph of cross-sections of human sperm flagella with PCD. In the bottom, a principal piece is visible with 9 + 2 axonemal pattern (Ax) devoid of dynein arms; FS: fibrous sheath. On the top, a cross-section of midpiece shows mitochondria (M), outer dense fibers (ODF), and a 9 + 2 axoneme devoid of dynein arms. Bar: 0.5 μm

spermatogenesis is poorly understood. Most of these genes encode proteins of ODA and IDA, dynein regulatory complexes, axonemal organization, and some of them are in common with MMAF (Jiao et al. 2021; Sironen et al. 2020). An estimated 70% of individuals with a clinical PCD phenotype are thought to have a bi-allelic mutation in a gene known to cause PCD (Shapiro et al. 2018).

Jiao et al. (2021) reported several genes found mutated in PCD linked to male infertility: dynein axonemal assembly factor or *DNAAF* involved in the preassembly and stability of ODA and IDA, DNAAF interacting proteins, radial spoke components, members of heat shock proteins. In addition, coiled-coil domain-containing CCDC39 and CCDC40 mutations are detected in PCD patients with ultrastructural defects in IDAs and axonemal organization (Jiao et al. 2021).

5.5.3 Dysplasia of the Fibrous Sheath/Multiple Morphological Abnormalities of the Sperm Flagella

Dysplasia of the fibrous sheath (DFS)/ MMAF is a systematic defect with genetic basis and familial incidence that compromises flagellar cytoskeletal components. Due to the peculiar alterations of the tails, the defect was named in the past as "stump tail," "short tail" (Baccetti et al. 1993), and DFS (Chemes et al. 1987), then, recently, MMAF (Ben Khelifa et al. 2014). These different denominations reflect the fact that this defect can be observed in complete and incomplete form and constitute a group of interrelated cytoskeletal pathologies affecting respiratory cilia and/or sperm flagella. Here, the defect will be identified as DFS/MMAF.

The head of DFS/MMAF spermatozoa is generally normal, and the alteration seems to occur during flagellum elongation in spermatids at the latest stages of spermiogenesis. At LM level and SEM (Fig. 5.4), the most evident sperm phenotype appears as short, thick, irregular tails and absent midpieces in high percentage.

The ultrastructural study is advisable to distinguish DFS/MMAF from NFSA and tail breaking secondary to sperm death and necrosis. The main evident ultrastructural defect is the dysplastic fibrous sheath components that form thick multilayered rings around axonemes often with disorganized microtubules. Axonemes were rarely normal, frequently lacked central pairs or peripheral doublets or were completely disorganized. Dynein arms were either present or missing. Respiratory involvement was documented in approximately 20% of the patients (Chemes 2017, 2018). Very rare combinations of DFS/MMAF and alterations in head-tail attachment characterized by TEM have been also reported (Moretti et al. 2011, 2017; Rawe et al. 2002).

To date, natural conceptions in DFS/MMAF have been not reported and due to severe asthenozoospermia, ICSI is the ART technique used in these cases with promising results, even if several concerns regard the potential risk of transmitting mutations to offspring.

Although morphological features of the defect are known from several decades, the genetic basis of DFS/MMAF is only recently fully investigated.

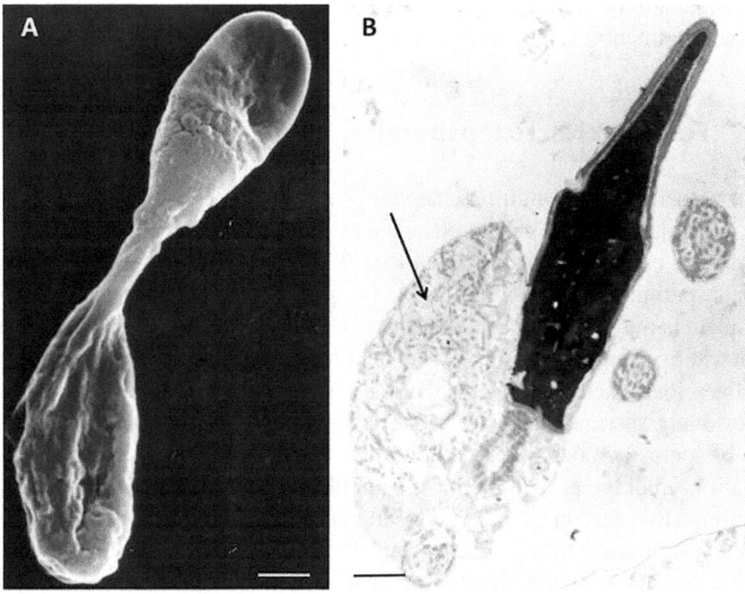

Fig. 5.4 Scanning (**A**) and transmission (**B**) electron microscopy micrograph of spermatozoa with DFS/MMAF. In (**A**), a typical short and thick tail embedded in a cytoplasmic remnant is shown. In (**B**) a longitudinal section of a spermatozoon with an irregular nucleus, an acrosome reduced in dimension and a cytoplasm with remnant of fibrous sheath (arrow). Bars: 1 μm

Two main structural proteins, AKAP3 and AKAP4, whose mutations were investigated in the early 2000s, compose fibrous sheath. Turner et al. (2001) did not detect alterations at the protein or gene levels in nine DFS/MMAF patients, whereas Baccetti et al. (2005b) demonstrated intragenic deletions of the *AKAP3* and *AKAP4* genes and absence of the AKAP4 protein in sperm flagella from a DFS patient. Then, research by Ben Khelifa et al. (2014) discovered *DNAH1,* whose mutations are the most common identified cause of DSF/MMAF. *DNAH1* encodes for an axonemal dynein of the IDA expressed in the testis. Further genetic investigations contribute to the identification of mutated genes involved in DFS/MMAF. Indeed, in the past five years, 40 genes whose mutations cause DFS/MMAF were identified and reported in several review papers (Martinez et al. 2023; Touré et al. 2021; Wang et al. 2020). The repertory of mutations inducing DFS/MMAF affects genes encoding IDA/ODA complexes (*DNAH1, DNAH2, DNAH6, DNAH17,* and others), radial spoke complexes (*CFAP251*), periaxonemal structure (*FSIP2),* centrosome (*CEP135*), intraflagellar transport complex (*TTC21A, TTC29, SPEF2, CFAP69*), and many others (Touré et al. 2021).

The research on genetic mutations related DFS/MMAF are extremely dynamic and many papers on this topic from the last 2 years are present in the literature (Jreijiri et al. 2024; Kherraf et al. 2024; Liu et al. 2023b, 2024; Martinez et al. 2023; Muroňová et al. 2024; Shao et al. 2024 and others).

The use of current understanding to identify opportunities and priorities for future research in the field of genetic causes of infertility is a relevant issue, as well as the application of mutation knowledge to risk prediction, genetic diagnosis, and potential treatment for human infertility.

5.6 Focus on Necrozoospermia

Necrozoospermia is a condition characterized by high percentage of dead spermatozoa in an ejaculate and can be observed as mild, moderate, and severe necrozoospermia. It is estimated that only 0.2% to 0.5% of infertile men suffer from complete necrozoospermia (Brahem et al. 2012). The parameter useful for necrozoospermia assessment is the sperm vitality, a measure of the percentage of living spermatozoa. Sperm vitality evaluation is included in the World Health Organization (WHO) guidelines for semen analysis that, in the last edition, reported the lower limit to 54% of living spermatozoa (WHO 2021). Necrozoospermia can occur at different levels of sperm production, essentially defective spermatogenesis, seminal plasma anomalies, epididymal dysfunctions, testicular hyperthermia, blood–testis barrier alterations (Boursier et al. 2022). Necrozoospermia is strongly associated with sperm DNA fragmentation (Brahem et al. 2012; Zhang et al. 2024) which can reduce the chances of spontaneous pregnancy or affect the outcome of ART. The ultrastructural characteristics of spermatozoa in a case of total necrozoospermia include broken plasma membrane, reacted or absent acrosome, nucleus, often irregularly shaped, chromatin with a disrupted texture, swollen mitochondria, and disorganized mitochondrial helix, altered axonemal and periaxonemal structures (Moretti et al. 2006).

The causes and risk factors for necrozoospermia include genital tract infections, that are the most common cause, testicular hyperthermia (local or systemic as fever), varicocele, hyperthyroidism, spinal cord injury, polycystic kidney disease, antisperm antibodies, advanced male age, toxic substances as tobacco, cannabis, and pesticides; a consistent percentage of necrozoospermia remains undiagnosed and defined as idiopathic (Agarwal et al. 2022).

Sperm viability may be assessed using several techniques, including the hypoosmotic swelling test and the eosin-nigrosin stain. Hypoosmotic swelling test is based on the fact that cells with intact membrane tend to reach the osmotic equilibrium. If the spermatozoa with intact membrane are placed in a hypoosmotic solution, the water cross the membrane, causing a very reproducible swelling of the sperm tails. In dead sperm, no changes in flagellar morphology occur in the presence of the media (Sanyal et al. 2024).

The eosin-nigrosin stain is a test based on dye exclusion: in the presence of eosin, dead spermatozoa appear red or dark pink while live spermatozoa remain unstained (Mortimer 2020).

Sperm vitality evaluation is mostly advised if the overall motility is less than 40%, according to the WHO Laboratory Manual's sixth edition (2021). Therefore, a motile spermatozoon is considered alive, and in specific conditions an immotile spermatozoon can also be alive. Thus, in order to guide future research and treatment of infertile individuals, it is crucial to distinguish between asthenozoospermia, which is a pathological decrease in sperm motility, and necrozoospermia, which is a pathological decrease in sperm vitality (Agarwal et al. 2022).

If available, an etiological treatment of necrozoospermia should be performed depending on the cause, if known. In addition, several studies reported the efficacy of repeated ejaculations in improving sperm vitality. Some drug treatments also seem to improve sperm vitality, although official guidelines encouraging their prescription are not available. Finally, in the presence of severe necrozoospermia, a chance of successful ICSI procedure could be the use of surgically extracted testicular spermatozoa (Boursier et al. 2022).

5.7 Conclusion

Humans exhibit a notably high percentage of sperm abnormalities, which can be observed even in healthy and fertile individuals. The study of alterations affecting human spermatozoa is far from being fully exhaustive and requires further investigations. From a clinical perspective, in case of systematic sperm abnormalities, the defective morphology of spermatozoa compromises the natural fertilization process and can also cause problems during the assisted fertilization procedures. The study of patients with systematic sperm defects such as globozoospermia, macrozoospermia, dysplasia of fibrous sheath, and others allowed the identification of several key genes involved in spermatogenesis. A more challenging task is to delineate the impact of non-systematic sperm abnormalities in the fertilization process because they are present in variable percentage among individuals and in different ejaculates of the same subject. We should conclude by emphasizing that an accurate diagnosis of sperm abnormalities, which identifies the defective mechanisms involved, is the starting point for an evidence-based approach to treat severe male factor infertility.

Acknowledgments We would like to express our extreme gratitude to Dr. Laura Liguori for her support in creating images used in this chapter.

References

Afzelius BA, Eliasson R, Johnsen O, Lindholmer C (1975) Lack of dynein arms in immotile human spermatozoa. J Cell Biol 66(2):225–232. https://doi.org/10.1083/jcb.66.2.225

Agarwal A, Sharma RK, Gupta S, Boitrelle F, Finelli R, Parekh N, Durairajanayagam D, Saleh R, Arafa M, Cho CL, Farkouh A, Rambhatla A, Henkel R, Vogiatzi P, Tadros N, Kavoussi P, Ko E, Leisegang K, Kandil H, Palani A, Salvio G, Mostafa T, Rajmil O, Banihani SA, Schon S, Le TV, Birowo P, Çeker G, Alvarez J, Molina JMC, Ho CCK, Calogero AE, Khalafalla K, Duran MB, Kuroda S, Colpi GM, Zini A, Anagnostopoulou C, Pescatori E, Chung E, Caroppo E, Dimitriadis F, Pinggera GM, Busetto GM, Balercia G, Elbardisi H, Taniguchi H, Park HJ, Maldonado Rosas I, de la Rosette J, Ramsay J, Bowa K, Simopoulou M, Rodriguez MG, Sabbaghian M, Martinez M, Gilani MAS, Al-Marhoon MS, Kosgi R, Cannarella R, Micic S, Fukuhara S, Parekattil S, Jindal S, Abdel-Meguid TA, Morimoto Y, Shah R (2022) Sperm vitality and Necrozoospermia: diagnosis, management, and results of a global survey of clinical practice. World J Mens Health 40(2):228–242. https://doi.org/10.5534/wjmh.210149

Auger J, Jouannet P, Eustache F (2016) Another look at human sperm morphology. Hum Reprod 31(1):10–23. https://doi.org/10.1093/humrep/dev251

Avidor-Reiss T, Carr A, Fishman EL (2020) The sperm centrioles. Mol Cell Endocrinol 518:110987. https://doi.org/10.1016/j.mce.2020.110987

Baccetti B, Fraioli F, Paolucci D, Selmi G, Spera G, Renieri T (1978) Double spermatozoa in a hyperprolactinemic man. J Submicrosc Cytol Pathol 10:240–260

Baccetti B, Burrini AG, Collodel G, Magnano AR, Piomboni P, Renieri T, Sensini C (1989) Crater defect in human spermatozoa. Gamete Res 22(3):249–255. https://doi.org/10.1002/mrd.1120220302

Baccetti B, Burrini AG, Collodel G, Piomboni P, Renieri T (1991) A "miniacrosome" sperm defect causing infertility in two brothers. J Androl 12(2):104–111

Baccetti B, Burrini AG, Capitani S, Collodel G, Moretti E, Piomboni P, Renieri T (1993) Notulae seminologicae. 2. The 'short tail' and 'stump' defect in human spermatozoa. Andrologia 25(6):331–335

Baccetti B, Capitani S, Collodel G, Di Cairano G, Gambera L, Moretti E, Piomboni P (2001) Genetic sperm defects and consanguinity. Hum Reprod 16:1365–1371. https://doi.org/10.1093/humrep/16.7.1365

Baccetti B, Capitani S, Collodel G, Strehler E, Piomboni P (2002) Recent advances in human sperm pathology. Contraception 65(4):283–287. https://doi.org/10.1016/s0010-7824(02)00290-1

Baccetti B, Collodel G, Marzella R, Moretti E, Piomboni P, Scapigliati G, Serafini F (2005a) Ultrastructural studies of spermatozoa from infertile males with Robertsonian translocations and 18, X, Y aneuploidies. Hum Reprod 20:2295–2300

Baccetti B, Collodel G, Estenoz M, Manca D, Moretti E, Piomboni P (2005b) Gene deletions in an infertile man with sperm fibrous sheath dysplasia. Hum Reprod 20(10):2790–2794. https://doi.org/10.1093/humrep/dei126

Ben Khelifa M, Coutton C, Zouari R, Karaouzène T, Rendu J, Bidart M, Yassine S, Pierre V, Delaroche J, Hennebicq S, Grunwald D, Escalier D, Pernet-Gallay K, Jouk PS, Thierry-Mieg N, Touré A, Arnoult C, Ray PF (2014) Mutations in DNAH1, which encodes an inner arm heavy chain dynein, lead to male infertility from multiple morphological abnormalities of the sperm flagella. Am J Hum Genet 94(1):95–104. https://doi.org/10.1016/j.ajhg.2013.11.017

Berkovitz A, Eltes F, Ellenbogen A, Peer S, Feldberg D, Bartoov B (2006) Does the presence of nuclear vacuoles in human sperm selected for ICSI affect pregnancy outcome? Hum Reprod 21(7):1787–1790. https://doi.org/10.1093/humrep/del049

Beurois J, Cazin C, Kherraf ZE, Martinez G, Celse T, Touré A, Arnoult C, Ray PF, Coutton C (2020) Genetics of teratozoospermia: Back to the head. Best Pract Res Clin Endocrinol Metab 34(6):101473. https://doi.org/10.1016/j.beem.2020.101473

Boguenet M, Bouet PE, Spiers A, Reynier P, May-Panloup P (2021) Mitochondria: their role in spermatozoa and in male infertility. Hum Reprod Update 27(4):697–719. https://doi.org/10.1093/humupd/dmab001

Boitrelle F, Ferfouri F, Petit JM, Segretain D, Tourain C, Bergere M, Bailly M, Vialard F, Albert M, Selva J (2011) Large human sperm vacuoles observed in motile spermatozoa under high magnification: nuclear thumbprints linked to failure of chromatin condensation. Hum Reprod 26:1650–1658

Boursier A, Dumont A, Boitrelle F, Prasivoravong J, Lefebvre-Khalil V, Robin G, Barbotin AL (2022) Necrozoospermia: the tree that hides the forest. Andrology 10(4):642–659. https://doi.org/10.1111/andr.13172

Brahem S, Jellad S, Ibala S, Saad A, Mehdi M (2012) DNA fragmentation status in patients with necrozoospermia. Syst Biol Reprod Med 58(6):319–323. https://doi.org/10.3109/19396368.2012.710869

Cannarella R, Maniscalchi ET, Condorelli RA, Scalia M, Guerri G, La Vignera S, Bertelli M, Calogero AE (2020) Ultrastructural sperm flagellum defects in a patient with CCDC39 compound heterozygous mutations and primary ciliary dyskinesia/situs Viscerum inversus. Front Genet 11:974. https://doi.org/10.3389/fgene.2020.00974

Chemes HE (2000) Phenotypes of sperm pathology: genetic and acquired forms in infertile men. J Androl 21:799–808

Chemes HE (2017) Sperm ultrastructure in fertile men and male sterility: revisiting Teratozoospermia. In: De Jonge CJ, Barratt CLR (eds) The sperm cell: production, maturation, fertilization, regeneration. Cambridge University Press, pp 36–58

Chemes HE (2018) Phenotypic varieties of sperm pathology: genetic abnormalities or environmental influences can result in different patterns of abnormal spermatozoa. Anim Reprod Sci 194:41–56. https://doi.org/10.1016/j.anireprosci.2018.04.074

Chemes HE, Alvarez Sedo C (2012) Tales of the tail and sperm head aches: changing concepts on the prognostic significance of sperm pathologies affecting the head, neck and tail. Asian J Androl 14:14–23. https://doi.org/10.1038/aja.2011.168

Chemes HE, Rawe YV (2003) Sperm pathology: a step beyond descriptive morphology. Origin, characterization and fertility potential of abnormal sperm phenotypes in infertile men. Hum Reprod Update 9:405–428. https://doi.org/10.1093/humupd/dmg034

Chemes HE, Brugo S, Zanchetti F, Carrere C, Lavieri JC (1987) Dysplasia of the fibrous sheath: an ultrastructural defect of human spermatozoa associated with sperm immotility and primary sterility. Fertil Steril 48(4):664–669. https://doi.org/10.1016/s0015-0282(16)59482-5

Chemes HE, Morero JL, Lavieri JC (1990) Extreme asthenozoospermia and chronic respiratory disease: a new variant of the immotile cilia syndrome. Int J Androl 13:216–222

Chianese R, Pierantoni R (2021) Mitochondrial reactive oxygen species (ROS) production alters sperm quality. Antioxidants (Basel) 10(1):92. https://doi.org/10.3390/antiox10010092

Collodel G, Iacoponi F, Mazzi L, Terzuoli G, Pascarelli NA, Moretti E (2013) Light, polarizing, and transmission electron microscopy: three methods for the evaluation of sperm quality. Syst Biol Reprod Med 59(1):27–33. https://doi.org/10.3109/19396368.2012.724518

Coudert A, Cazin C, Amiri-Yekta A, Fourati Ben Mustapha S, Zouari R, Bessonat J, Zoghmar A, Clergeau A, Metzler-Guillemain C, Triki C, Lejeune H, Sermondade N, Pipiras E, Prisant

N, Cedrin I, Koscinski I, Keskes L, Lestrade F, Hesters L, Rives N, Dorphin B, Guichet A, Patrat C, Dulioust E, Feraille A, Robert F, Brouillet S, Morel F, Perrin A, Rougier N, Bieth E, Sorlin A, Siffroi JP, Ben Khelifa M, Boiterelle F, Hennebicq S, Satre V, Arnoult C, Coutton C, Barbotin AL, Thierry-Mieg N, Kherraf ZE, Ray PF (2023) Genetic causes of macrozoospermia and proposal for an optimized genetic diagnosis strategy based on sperm parameters. J Genet Genomics 50(7):536–540. https://doi.org/10.1016/j.jgg.2023.04.007

Coutton C, Escoffier J, Martinez G, Arnoult C, Ray PF (2015) Teratozoospermia: spotlight on the main genetic actors in the human. Hum Reprod Update 21(4):455–485. https://doi.org/10.1093/humupd/dmv020

Crafa A, Condorelli RA, La Vignera S, Calogero AE, Cannarella R (2023) Globozoospermia: a case report and systematic review of literature. World J Mens Health 41(1):49–80. https://doi.org/10.5534/wjmh.220020

De Braekeleer M, Nguyen MH, Morel F, Perrin A (2015) Genetic aspects of monomorphic teratozoospermia: a review. J Assist Reprod Genet 32(4):615–623. https://doi.org/10.1007/s10815-015-0433-2

Di Emidio G, Falone S, Artini PG, Amicarelli F, D'Alessandro AM, Tatone C (2021) Mitochondrial sirtuins in reproduction. Antioxidants (Basel) 10(7):1047. https://doi.org/10.3390/antiox10071047

Elkhatib RA, Paci M, Longepied G, Saias-Magnan J, Courbière B, Guichaoua MR, Lévy N, Metzler-Guillemain C, Mitchell MJ (2017) Homozygous deletion of SUN5 in three men with decapitated spermatozoa. Hum Mol Genet 26(16):3167–3171. https://doi.org/10.1093/hmg/ddx200

Escalier D (2003) New insights into the assembly of the periaxonemal structures in mammalian spermatozoa. Biol Reprod 69:373–378

Escalier D, David G (1984) Pathology of the cytoskeleton of the human sperm flagellum: axonemal and peri-axonemal anomalies. Biol Cell 50:37–52

Fang F, Ni K, Shang J, Zhang X, Xiong C, Meng T (2018) Expression of mitofusin 2 in human sperm and its relationship to sperm motility and cryoprotective potentials. Exp Biol Med (Maywood) 243:963–969

Fellmeth JE, Ghanaim EM, Schindler K (2016) Characterization of macrozoospermia-associated AURKC mutations in a mammalian meiotic system. Hum Mol Genet 25:2698–2711. https://doi.org/10.1093/hmg/ddw128

Ferramosca A, Provenzano SP, Coppola L, Zara V (2012) Mitochondrial respiratory efficiency is positively correlated with human sperm motility. Urology 79:809–814

Fesahat F, Henkel R, Agarwal A (2020) Globozoospermia syndrome: an update. Andrologia 52:e13459

Fishman EL, Jo K, Nguyen QPH, Kong D, Royfman R, Cekic AR, Khanal S, Miller AL, Simerly C, Schatten G, Loncarek J, Mennella V, Avidor-Reiss T (2018) A novel atypical sperm centriole is functional during human fertilization. Nat Commun 9:2210. https://doi.org/10.1038/s41467-018-04678-8

Franco JG Jr, Mauri AL, Petersen CG, Massaro FC, Silva LF, Felipe V, Cavagna M, Pontes A, Baruffi RL, Oliveira JB, Vagnini LD (2012) Large nuclear vacuoles are indicative of abnormal chromatin packaging in human spermatozoa. Int J Androl 35:46–51

Galletta BJ, Ortega JM, Smith SL, Fagerstrom CJ, Fear JM, Mahadevaraju S, Oliver B, Rusan NM (2020) Sperm head-tail linkage requires restriction of pericentriolar material to the proximal centriole end. Dev Cell 53(1):86–101.e7. https://doi.org/10.1016/j.devcel.2020.02.006

Gallon F, Marchetti C, Jouy N, Marchetti P (2006) The functionality of mitochondria differentiates human spermatozoa with high and low fertilizing capability. Fertil Steril 86:1526–1530

Garanina AS, Alieva IB, Bragina EE, Blanchard E, Arbeille B, Guerif F, Uzbekova S, Uzbekov RE (2019) The centriolar adjunct⁻Appearance and disassembly in spermiogenesis and the potential impact on fertility. Cells 8(2):180. https://doi.org/10.3390/cells8020180

Gómez-Torres MJ, Luna-Romero J, Fernández-Colom PJ, Aizpurua J, Avilés M, Romero A (2023) Human sperm head vacuoles are related to nuclear-envelope invaginations. Int J Mol Sci 24(12):10027. https://doi.org/10.3390/ijms241210027

Grunewald S, Fitzl G, Springsguth C (2017) Induction of ultra-morphological features of apoptosis in mature and immature sperm. Asian J Androl 19(5):533–537. https://doi.org/10.4103/1008-682X.180974

Gu NH, Zhao WL, Wang GS, Sun F (2019) Comparative analysis of mammalian sperm ultrastructure reveals relationships between sperm morphology, mitochondrial functions and motility. Reprod Biol Endocrinol 17(1):66. https://doi.org/10.1186/s12958-019-0510-y

Jaiswal A, Baliu-Souza T, Turner K, Nadiminty N, Rambhatla A, Agarwal A, Krawetz SA, Dupree JM, Saltzman B, Schon SB, Avidor-Reiss T (2022) Sperm centriole assessment identifies male factor infertility in couples with unexplained infertility - a pilot study. Eur J Cell Biol 101(3):151243. https://doi.org/10.1016/j.ejcb.2022.151243

Jiao SY, Yang YH, Chen SR (2021) Molecular genetics of infertility: loss-of-function mutations in humans and corresponding knockout/mutated mice. Hum Reprod Update 27(1):154–189. https://doi.org/10.1093/humupd/dmaa034

Jreijiri F, Cavarocchi E, Amiri-Yekta A, Cazin C, Hosseini SH, El Khouri E, Patrat C, Thierry-Mieg N, Ray PF, Dulioust E, Whitfield M, Touré A (2024) CCDC65, encoding a component of the axonemal nexin-dynein regulatory complex, is required for sperm flagellum structure in humans. Clin Genet 105(3):317–322. https://doi.org/10.1111/cge.14459

Kherraf ZE, Barbotin AL, Martinez G, Mazet A, Cazin C, Coutton C, Arnoult C, Thierry-Mieg N, Rives N, Rives-Feraille A, Ray PF (2024) A splice donor variant of GAS8 induces structural disorganization of the axoneme in sperm flagella and leads to nonsyndromic male infertility. Clin Genet 105(2):220–225. https://doi.org/10.1111/cge.14450

Krausz C, Riera-Escamilla A (2018) Genetics of male infertility. Nat Rev Urol 15(6):369–384. https://doi.org/10.1038/s41585-018-0003-3

Lhuillier P, Rode B, Escalier D, Lorès P, Dirami T, Bienvenu T, Gacon G, Dulioust E, Touré A (2009) Absence of annulus in human asthenozoospermia: case report. Hum Reprod 24(6):1296–1303. https://doi.org/10.1093/humrep/dep020

Li L, Sha Y, Wang X, Li P, Wang J, Kee K, Wang B (2017) Whole-exome sequencing identified a homozygous BRDT mutation in a patient with acephalic spermatozoa. Oncotarget 8(12):19914–19922. https://doi.org/10.18632/oncotarget.15251

Liu G, Xing X, Zhang H, Zhu W, Lin G, Lu G, Li W (2021) Patients with acephalic spermatozoa syndrome linked to novel TSGA10/PMFBP1 variants have favorable pregnancy outcomes from intracytoplasmic sperm injection. Clin Genet 100(3):334–339. https://doi.org/10.1111/cge.14007

Liu Y, Li Y, Meng L, Li K, Gao Y, Lv M, Guo R, Xu Y, Zhou P, Wei Z, He X, Cao Y, Wu H, Tan Y, Hua R (2023a) Bi-allelic human TEKT3 mutations cause male infertility with oligoasthenoteratozoospermia owing to acrosomal hypoplasia and reduced progressive motility. Hum Mol Genet 32(10):1730–1740. https://doi.org/10.1093/hmg/ddad013

Liu M, Dai S, Zhang J, Yang Y, Shen Y, Liu H, Yang Y, Jiang C, Tian E (2023b) A novel mutation in CFAP47 causes male infertility due to multiple morphological abnormalities of the sperm flagella. Front Endocrinol (Lausanne) 14:1155639. https://doi.org/10.3389/fendo.2023.1155639

Liu Z, Wang C, Ni F, Li T, Yang F, Wei H, Li T, Huang C, Wang J, Wang B (2024) Identification of a homozygous mutation of CCDC40 in a Chinese infertile man with MMAF and PCD-like phenotypes. Genet Test Mol Biomarkers 28(8):337–341. https://doi.org/10.1089/gtmb.2023.0263

Martinez G, Barbotin AL, Cazin C, Wehbe Z, Boursier A, Amiri-Yekta A, Daneshipour A, Hosseini SH, Rives N, Feraille A, Thierry-Mieg N, Bidart M, Satre V, Arnoult C, Ray PF, Kherraf ZE, Coutton C (2023) New mutations in DNHD1 cause multiple morphological abnormalities of the sperm flagella. Int J Mol Sci 24(3):2559. https://doi.org/10.3390/ijms24032559

Mazaheri Moghaddam M, Mazaheri Moghaddam M, Hamzeiy H, Baghbanzadeh A, Pashazadeh F, Sakhinia E (2021) Genetic basis of acephalic spermatozoa syndrome, and intracytoplasmic sperm injection outcomes in infertile men: a systematic scoping review. J Assist Reprod Genet 38(3):573–586. https://doi.org/10.1007/s10815-020-02008-w

Mitchell V, Sigala J, Ballot C, Jumeau F, Barbotin AL, Duhamel A, Rives N, Rigot JM, Escalier D, Peers MC (2015) Light microscopy morphological characteristics of the sperm flagellum

may be related to axonemal abnormalities. Andrologia 47(2):214–220. https://doi.org/10.1111/and.12249
Moretti E, Baccetti B, Scapigliati G, Collodel G (2006) Transmission electron microscopy, immunocytochemical and fluorescence in situ hybridisation studies in a case of 100% necrozoospermia: case report. Andrologia 38(6):233–238. https://doi.org/10.1111/j.1439-0272.2006.00741.x
Moretti E, Pascarelli NA, Federico MG, Renieri T, Collodel G (2008) Abnormal elongation of midpiece, absence of axoneme and outer dense fibers at principal piece level, supernumerary microtubules: a sperm defect of possible genetic origin? Fertil Steril 90(4):1201.e3–1201.e8. https://doi.org/10.1016/j.fertnstert.2007
Moretti E, Pascarelli NA, Giannerini V, Geminiani M, Anichini C, Collodel G (2009) 18, X, Y aneuploidies and transmission electron microscopy studies in spermatozoa from five carriers of different reciprocal translocations. Asian J Androl 11:325–332
Moretti E, Geminiani M, Terzuoli G, Renieri T, Pascarelli N, Collodel G (2011) Two cases of sperm immotility: a mosaic of flagellar alterations related to dysplasia of the fibrous sheath and abnormalities of head-neck attachment. Fertil Steril 95:1787.e19–1787.e23. https://doi.org/10.1016/j.fertnstert.2010.11.027
Moretti E, Sutera G, Collodel G (2016) The importance of transmission electron microscopy analysis of spermatozoa: diagnostic applications and basic research. Syst Biol Reprod Med 62:171–183. https://doi.org/10.3109/19396368.2016.1155242
Moretti E, Pascarelli NA, Belmonte G, Renieri T, Collodel G (2017) Sperm with fibrous sheath dysplasia and anomalies in head-neck junction: focus on centriole and centrin 1. Andrologia 49(7). https://doi.org/10.1111/and.12701
Moretti E, Collodel G, Salvatici MC, Belmonte G, Signorini C (2019) New insights into sperm with total globozoospermia: increased fatty acid oxidation and centrin1 alteration. Syst Biol Reprod Med 65(5):390–399. https://doi.org/10.1080/19396368.2019.1626934
Moretti E, Noto D, Guazzo R, Menchiari A, Belmonte G, Collodel G (2021) Centriolar defects, centrin 1 alterations, and FISH studies in human spermatozoa of a male partner of a couple that produces aneuploid embryos in natural and artificial fertilization. J Assist Reprod Genet 38(5):1197–1205. https://doi.org/10.1007/s10815-021-02109-0
Moretti E, Signorini C, Noto D, Corsaro R, Collodel G (2022) The relevance of sperm morphology in male infertility. Front Reprod Health 4:945351. https://doi.org/10.3389/frph.2022.945351
Mortimer D (2018) The functional anatomy of the human spermatozoon: relating ultrastructure and function. Mol Hum Reprod 24(12):567–592. https://doi.org/10.1093/molehr/gay040
Mortimer D (2020) A technical note on the assessment of human sperm vitality using eosin-nigrosin staining. Reprod Biomed Online 40(6):851–855. https://doi.org/10.1016/j.rbmo.2020.03.002
Mundy AJ, Ryder TA, Edmonds DK (1995) Asthenozoospermia and the human sperm mid-piece. Hum Reprod 10(1):116–119. https://doi.org/10.1093/humrep/10.1.116
Muroňová J, Kherraf ZE, Giordani E, Lambert E, Eckert S, Cazin C, Amiri-Yekta A, Court M, Chevalier G, Martinez G, Neirijnck Y, Kühne F, Wehrli L, Klena N, Hamel V, De Macedo L, Escoffier J, Guichard P, Coutton C, Mustapha SFB, Kharouf M, Bouin AP, Zouari R, Thierry-Mieg N, Nef S, Geimer S, Loeuillet C, Ray PF, Arnoult C (2024) Lack of CCDC146, a ubiquitous centriole and microtubule-associated protein, leads to non-syndromic male infertility in human and mouse. elife 12:RP86845. https://doi.org/10.7554/eLife.86845
Nie H, Tang Y, Qin W (2020) Beyond Acephalic spermatozoa: the complexity of intracytoplasmic sperm injection outcomes. Biomed Res Int 2020:6279795. https://doi.org/10.1155/2020/6279795
Oliveira JB, Massaro FC, Baruffi RL, Mauri AL, Petersen CG, Silva LF, Vagnini LD, Franco JG Jr (2010) Correlation between semen analysis by motile sperm organelle morphology examination and sperm DNA damage. Fertil Steril 94:1937–1940
Pastuszek E, Kiewisz J, Skowronska P, Liss J, Lukaszuk M, Bruszczynska A, Jakiel G, Lukaszuk K (2017) An investigation of the potential effect of sperm nuclear vacuoles in human spermatozoa on DNA fragmentation using a neutral and alkaline comet assay. Andrology 5(2):392–398. https://doi.org/10.1111/andr.12324

Pedersen H, Rebbe H (1975) Absence of arms in the axoneme of immobile human spermatozoa. Biol Reprod 12(5):541–544. https://doi.org/10.1095/biolreprod12.5.541

Perdrix A, Travers A, Chelli MH, Escalier D, Do Rego JL, Milazzo JP, Mousset-Siméon N, Macé B, Rives N (2011) Assessment of acrosome and nuclear abnormalities in human spermatozoa with large vacuoles. Hum Reprod 26:47–58

Rawe VY, Terada Y, Nakamura S, Chillik CF, Olmedo SB, Chemes HE (2002) A pathology of the sperm centriole responsible for defective sperm aster formation, syngamy and cleavage. Hum Reprod 17(9):2344–2349. https://doi.org/10.1093/humrep/17.9.2344

Rawe VY, Hermes R, Nodar FN, Fiszbajn G, Chemes HE (2007) Results of intracytoplasmic sperm injection in two infertile patients with abnormal organization of sperm mitochondrial sheaths and severe asthenoteratozoospermia. Fertil Steril 88(3):649–653. https://doi.org/10.1016/j.fertnstert.2006.12.074

Ray PF, Toure A, Metzler-Guillemain C, Mitchell MJ, Arnoult C, Coutton C (2017) Genetic abnormalities leading to qualitative defects of sperm morphology or function. Clin Genet 91(2):217–232. https://doi.org/10.1111/cge.12905

Ribas-Maynou J, Nguyen H, Wu H, Ward WS (2022) Functional aspects of sperm chromatin organization. Results Probl Cell Differ 70:295–311. https://doi.org/10.1007/978-3-031-06573-6_10

Ricci G, Andolfi L, Zabucchi G, Luppi S, Boscolo R, Martinelli M, Zweyer M, Trevisan E (2015) Ultrastructural morphology of sperm from human Globozoospermia. Biomed Res Int 2015:798754. https://doi.org/10.1155/2015/798754

Rosati AJ, Whitcomb BW, Brandon N, Buck Louis GM, Mumford SL, Schisterman EF, Pilsner JR (2020) Sperm mitochondrial DNA biomarkers and couple fecundity. Hum Reprod 35:deaa191–2625

Sanyal D, Arya D, Nishi K, Balasinor N, Singh D (2024) Clinical utility of sperm function tests in predicting male fertility: a systematic review. Reprod Sci 31(4):863–882. https://doi.org/10.1007/s43032-023-01405-7

Sha YW, Wang X, Xu X, Ding L, Liu WS, Li P, Su ZY, Chen J, Mei LB, Zheng LK, Wang HL, Kong SB, You M, Wu JF (2019) Biallelic mutations in PMFBP1 cause acephalic spermatozoa. Clin Genet 95(2):277–286. https://doi.org/10.1111/cge.13461

Sha Y, Wang X, Yuan J, Zhu X, Su Z, Zhang X, Xu X, Wei X (2020) Loss-of-function mutations in centrosomal protein 112 is associated with human acephalic spermatozoa phenotype. Clin Genet 97(2):321–328. https://doi.org/10.1111/cge.13662

Shao ZM, Zhu YT, Gu M, Guo SC, Yu H, Li KK, Tang DD, Xu YP, Lv MR (2024) Novel variants in DNAH6 cause male infertility associated with multiple morphological abnormalities of the sperm flagella (MMAF) and ICSI outcomes. Asian J Androl 26(1):91–98. https://doi.org/10.4103/aja202328

Shapiro AJ, Davis SD, Polineni D, Manion M, Rosenfeld M, Dell SD, Chilvers MA, Ferkol TW, Zariwala MA, Sagel SD, Josephson M, Morgan L, Yilmaz O, Olivier KN, Milla C, Pittman JE, Daniels MLA, Jones MH, Janahi IA, Ware SM, Daniel SJ, Cooper ML, Nogee LM, Anton B, Eastvold T, Ehrne L, Guadagno E, Knowles MR, Leigh MW, Lavergne V (2018) American Thoracic Society assembly on Pediatrics diagnosis of primary ciliary dyskinesia. An official American Thoracic Society clinical practice guideline. Am J Respir Crit Care Med 197:e24–e39. https://doi.org/10.1164/rccm.201805-0819ST

Sironen A, Shoemark A, Patel M, Loebinger MR, Mitchison HM (2020) Sperm defects in primary ciliary dyskinesia and related causes of male infertility. Cell Mol Life Sci 77(11):2029–2048. https://doi.org/10.1007/s00018-019-03389-7

Tanaka A, Nagayoshi M, Tanaka I, Kusunoki H (2012) Human sperm head vacuoles are physiological structures formed during the sperm development and maturation process. Fertil Steril 98:315–320

Tesarik J, Mendoza-Tesarik R (2023) Mitochondria in human fertility and infertility. Int J Mol Sci 24(10):8950. https://doi.org/10.3390/ijms24108950

Touré A, Martinez G, Kherraf ZE, Cazin C, Beurois J, Arnoult C, Ray PF, Coutton C (2021) The genetic architecture of morphological abnormalities of the sperm tail. Hum Genet 140(1):21–42. https://doi.org/10.1007/s00439-020-02113-x

Toyama Y, Iwamoto T, Yajima M, Baba K, Yuasa S (2000) Decapitated and decaudated spermatozoa in man, and pathogenesis based on the ultrastructure. Int J Androl 23(2):109–115. https://doi.org/10.1046/j.1365-2605.2000.t01-1-00217.x

Turner RM, Musse MP, Mandal A, Klotz K, Jayes FC, Herr JC, Gerton GL, Moss SB, Chemes HE (2001) Molecular genetic analysis of two human sperm fibrous sheath proteins, AKAP4 and AKAP3, in men with dysplasia of the fibrous sheath. J Androl 22(2):302–315

Turner KA, Fishman EL, Asadullah M, Ott B, Dusza P, Shah TA, Sindhwani P, Nadiminty N, Molinari E, Patrizio P, Saltzman BS, Avidor-Reiss T (2021) Fluorescence-based Ratiometric analysis of sperm centrioles (FRAC) finds patient age and sperm morphology are associated with centriole quality. Front Cell Dev Biol 9:658891. https://doi.org/10.3389/fcell.2021.658891

Wang WL, Tu CF, Tan YQ (2020) Insight on multiple morphological abnormalities of sperm flagella in male infertility: what is new? Asian J Androl 22(3):236–245. https://doi.org/10.4103/aja.aja_53_19

Wang Y, Xiang MF, Zheng N, Cao YX, Zhu FX (2022a) Genetic pathogenesis of acephalic spermatozoa syndrome: past, present, and future. Asian J Androl 24(3):231–237. https://doi.org/10.4103/aja202198

Wang G, Zhu X, Gao Y, Lv M, Li K, Tang D, Wu H, Xu C, Geng H, Shen Q, Zha X, Duan Z, Zhang J, Hua R, Tao F, Zhou P, Wei Z, Cao Y, Guo R, He X (2022b) Biallelic loss-of-function mutations in SEPTIN4 (C17ORF47), encoding a conserved annulus protein, cause thin midpiece spermatozoa and male infertility in humans. Hum Mutat 43(12):2079–2090. https://doi.org/10.1002/humu.24475

Wang Y, Huang X, Sun G, Chen J, Wu B, Luo J, Tang S, Dai P, Zhang F, Li J, Wang L (2024) Coiled-coil domain-containing 38 is required for acrosome biogenesis and fibrous sheath assembly in mice. J Genet Genomics 51(4):407–418. https://doi.org/10.1016/j.jgg.2023.09.002

Wilton LJ, Temple-Smith PD, de Kretser DM (1992) Quantitative ultrastructural analysis of sperm tails reveals flagellar defects associated with persistent asthenozoospermia. Hum Reprod 7:510–516

World Health Organization (2021) WHO laboratory manual for the examination and processing of human semen, 6th edn

Yao R, Ito C, Natsume Y, Sugitani Y, Yamanaka H, Kuretake S, Yanagida K, Sato A, Toshimori K, Noda T (2002) Lack of acrosome formation in mice lacking a Golgi protein, GOPC. Proc Natl Acad Sci USA 99(17):11211–11216. https://doi.org/10.1073/pnas.162027899

Yassine S, Escoffier J, Martinez G, Coutton C, Karaouzène T, Zouari R, Ravanat JL, Metzler-Guillemain C, Lee HC, Fissore R, Hennebicq S, Ray PF, Arnoult C (2015) Dpy19l2-deficient globozoospermic sperm display altered genome packaging and DNA damage that compromises the initiation of embryo development. Mol Hum Reprod 21(2):169–185. https://doi.org/10.1093/molehr/gau099

Zhang EC, Yin XY, Peng ZY, Lai ZZ, Hu L, Peng YQ, Zhang H, Ming R, Lin G, Li WN (2024) Novel insights into necrozoospermia from a single-center study: reference ranges, possible etiology, and impact on male fertility. Asian J Androl 26(5):528–534. https://doi.org/10.4103/aja202422

Zhu F, Wang F, Yang X, Zhang J, Wu H, Zhang Z, Zhang Z, He X, Zhou P, Wei Z, Gecz J, Cao Y (2016) Biallelic SUN5 mutations cause autosomal-recessive acephalic spermatozoa syndrome. Am J Hum Genet 99(4):942–949. https://doi.org/10.1016/j.ajhg.2016.08.004

Post-testicular Sperm Maturation: The Epididymis and the Seminal Plasma

6

Contents

Core Messages

In an in-depth study of the spermatozoon, the understanding of human physiology and pathology of the sperm epididymal maturation cannot be ignored. In this chapter, the epididymal transit of spermatozoa and the role of seminal plasma markers in predicting sperm quality and fertilization outcome are discussed. In the last years, the interest in studying the role of seminal plasma in male infertility is emerging.

It is well established that routinely assessed semen parameters offer valuable insights into an individual's fertility status; however, they lack both diagnostic and prognostic power. In this context, the diagnostic potential of seminal plasma presents promising opportunities for the identification of novel biomarkers to enhance the diagnosis of male infertility and to better predict fertilization outcomes.

Abbreviations

AQPs	Aquaporins
ART	Assisted reproduction technologies
HSPA2	Heat shock-related 70 kDa protein 2

E. Moretti, G. Collodel, *The Human Spermatozoon*,
https://doi.org/10.1007/978-3-031-97868-5_6

MIF	Macrophage migration inhibitory factor
nGPx4	Nuclear glutathione peroxidase 4
PRDX6	Peroxiredoxin 6
PUFAs	Polyunsaturated fatty acids
RvD1	Resolvin D1
WHO	World Health Organization

6.1 The Epididymis and the Seminal Plasma

The mammalian spermatozoa, after the spermiogenesis and their release into the lumen of seminiferous tubules, are incapable of fertilizing in vivo, and one of the most important reasons is the lack of motility. Essentially, the spermatozoa become able to fertilize gradually, undergoing many post-testicular functional modifications. Then, the trip of the spermatozoon is not only characterized by the movement to reach the egg but is also a real acquisition of different experiences. If the ejaculated spermatozoa are placed in contact with the egg, they are not able to fertilize.

To become fertilizing competent, spermatozoa need to undergo two extra-testicular maturation processes: the first in the male reproductive tract includes the epididymal maturation and the second in the female one where spermatozoa undergo capacitation and are activated to fertilize or to die (Varner 2015).

Even though the topic of the book is the human spermatozoon, our understanding on epididymal transit is based on a wide variety of mammals, essentially rodent species. The study of epididymal functions in humans is poorly documented for obvious reasons, and it is mainly based on clinical observations of pathological conditions, such as agenesis or obstruction of the excurrent duct (Sullivan and Mieusset 2016; Sullivan et al. 2019).

The focus of this chapter is the analysis of the sperm epididymal transit and the relation between seminal plasma and sperm quality and/or fertilization outcome. In the last years, the interest in studying and understanding the role of seminal plasma in male infertility is emerging.

It is well known that traditional semen parameters offer some insight into an individual's fertility status; however, as a diagnostic tool, they present numerous limitations. In contrast, the diagnostic potential of seminal plasma holds promise for future advancements in identifying novel biomarkers for the diagnosis of male infertility.

6.1.1 Epididymal Organization and Histology

The epididymis represents the symbol of all vertebrate species with internal fertilization. The evolution of epididymis seems dependent on the reproductive strategies of the different mammalian species. For this reason, it is difficult to extrapolate biological observations based on a particular mammalian group to other mammalian species that have different reproductive strategy (Sullivan et al. 2019).

In general, the functions of epididymis in the animal kingdom include sperm storage and sperm maturation, with variable importance depending on the group of vertebrates. The epididymal environment promotes a severe quality control system to preserve spermatozoa from external stressors and distinguish faulty cells for elimination (Ali et al. 2023).

As reported above, the study of human epididymis is difficult for the retrieval of well-preserved samples to investigate histology, transcriptomic, and proteomics.

The epididymis is a long tubule connected to the seminiferous tubules by efferent ducts and rete testis, and giving rise to vas deferens, ejaculatory duct, and urethra (Fig. 6.1).

Three relevant anatomical regions, each of which with peculiar characteristics and functions, can be distinguished in human epididymis: caput (head), corpus (body), and cauda (tail) (Fig. 6.1). These regions are organized in distinct areas by the presence of *septae* formed by connective tissue that in humans are poorly differentiated (Sullivan and Mieusset 2016).

The proximal region of the epididymis is involved in sperm maturation, while the cauda plays a role as spermatozoa storage; this is particularly evident in laboratory rodents and in polygamous species to assure a sperm reservoir for multiple mating.

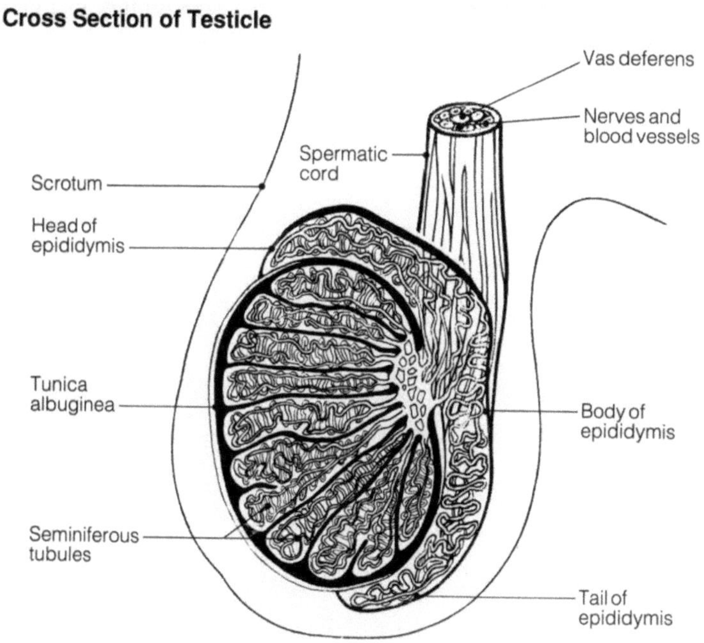

Fig. 6.1 Cross-section of the human testis and epididymis. The parenchima of the testis is divided in lobules by septa arising from the tunica albuginea. Each lobule contains seminiferous tubules. The epididymis is a single, convoluted and elongated tubule attached to the posterior side of the testis that connects the multiple efferent ducts to the vas deferens. Three main regions are present in the epididymis: the head (caput), the body (corpus) and the tail (cauda). Reprinted from National Cancer Institute (https://www.cancer.gov)

On the contrary, the epididymal cauda in humans has a limited sperm reservoir capacity and consequently this area is poorly developed. In humans, the characterization of the initial segment, formed by efferent ducts rich in ciliated cells is also difficult. This region is involved in estrogen-dependent water reabsorption and nuclear estrogen receptor alpha was found in efferent duct epithelial cells (Sullivan et al. 2019).

In mammalian species, the length of epididymis and the time required for sperm transit are reputed to be relatively long. The studies of transit duration of human spermatozoa along the epididymis were scant and gave contrasting results (Sullivan et al. 2019).

The regions of epididymis are characterized by a peculiar cells composition that leads to the gradual maturation of spermatozoa obtained in the different districts.

The cell types present in the epididymal duct epithelium are highly conserved in mammals. As a rule, the activity of these cells consists in a remodeling of the epididymal fluid to generate a luminal microenvironment suitable for sperm maturation. The epididymal epithelium is composed of six main cell types, some of which are distributed across the entire epididymis and others are restricted to particular areas. Androgens are the main regulators of the high metabolic, endocytic, and secretory activities of epididymal cells (James et al. 2020). It is noteworthy that the different cell categories show specific protein expression profiles that help characterize these cells.

Here, a brief description of the different cell types (Fig. 6.2) is provided.

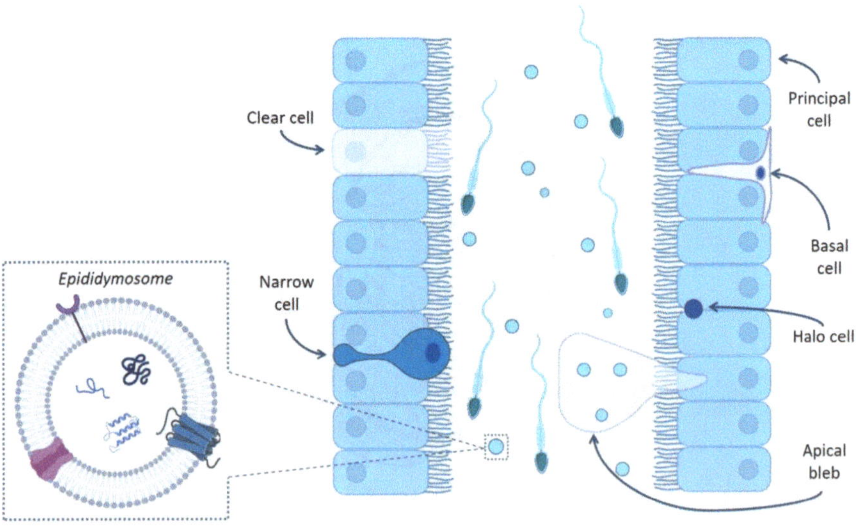

Fig. 6.2 Simplified representation of the internal structure of the epididymis. In the epithelium, different cells are visible as principal cells, clear cells, basal cells, halo cells, and narrow cells. In particular, principal cells producing epididymosomes (see Sect. 6.2) that contain different proteins, nucleic acids, and lipids. The inset represents a structure of an epididymosome, a vesicle containing different molecules

Principal Cells They are the main cell type in the epididymal epithelium and are distributed all along the duct. Depending on the region, the principal cells can represent up to 80% of the whole cell population. Although their principal biological activities change in the different epididymal regions, they are mainly responsible for absorption and secretion of materials into the epididymal lumen and therefore have high secretory and endocytic activities. They secrete different proteins in the luminal fluid and reabsorb others, modulating the protein composition of the fluid along the epididymal tract (Cornwall 2009). Principal cells are also the site of production and release of cargo-containing epididymosomes that are described later.

Apical Cells They are defined as typical cells of the initial segment of the epididymal epithelium. Their function remains relatively poorly understood, although significant endocytic activity has been observed in several species.

Narrow Cells They are typical cells of the initial segment. Although an endocytic activity was documented, their main role consists in the secretion of H^+ into the epididymal lumen.

Clear Cells They are observed exclusively within the caput, corpus, and cauda regions of the epididymis and are not present within the initial segment. They show high endocytic activity, greater than the other described cell types. The clear cells are also responsible for the removal of cytoplasmic droplets that are released from sperm cells during maturation in the epididymal lumen. Both narrow and clear cells express the vacuolar proton pump V-ATPase necessary for intraluminal acidification by active secretion of H^+ (Breton et al. 2016). In the human caput epididymis, V-ATPase-positive cells show morphological characteristics like narrow cells present in the initial segment of rodent epididymis.

Basal Cells As the principal cells, they are located all along the epididymal tubule (Fig. 6.2). They adhere to the basement membrane, and they do not have contact with epididymal lumen. These cells are an integral part of tubule structure, and it has been suggested that they may indirectly affect luminal environment by regulating some principal cell functions. In particular, they were purported to protect the male excurrent duct against reactive oxygen species, and to show endocrine functions, mainly the secretion of prostaglandins. Basal cells have also several features of adult stem cells, since they can differentiate in vitro and potentially regenerate epididymis epithelium.

Halo Cell They are present all along the epididymal epithelium and are the primary immune cells in the epididymis. They include macrophages/monocytes, mononuclear phagocytes, and T-lymphocytes, but not B-lymphocytes (Dufresne et al. 2024).

As is well known, the spermatozoa represent non-self-cells for the producing organism and therefore are highly immunogenic. Principal, narrow, clear, and basal cells form the blood–epididymis barrier by a series of different junctional complexes

(Sullivan and Belleannee 2018). This barrier is important in maintaining the composition of epididymal fluid different from the plasma and in protecting spermatozoa from immune system (Cyr et al. 2007). Possible damages occurring in the barrier can affect the physiological sperm maturation.

6.1.2 Functions of the Epididymis

As already mentioned, the transport of the spermatozoa from rete testis to vas deferens is accomplished in the epididymis with the smooth muscle layers that contract rhythmically and cilia present particularly in the cells of efferent ducts (Elfgen et al. 2018).

The epididymis also protects spermatozoa from damages caused by the external environment by several mechanisms. The cells of epididymal epithelium produce reactive oxygen species, harmful for spermatozoa, due to their high metabolic activity (O'Flaherty 2019). To protect spermatozoa and scavenge the reactive oxygen species, these cells secrete antioxidant enzymes into the lumen.

The antioxidant battery present in the epididymis includes superoxide dismutase, catalase, glutathione peroxidases, peroxiredoxins, glutathione-S-transferases, thioredoxins, and thioredoxin reductase (O'Flaherty 2019). The synergic action of these enzymes contributes to balance reactive oxygen species levels during epididymal maturation. Working with animal models gene deleted for the different enzymes, the importance of these antioxidants in controlling motility, chromatin compaction, and fertilization ability was evident.

The initial segment and the efferent ducts are areas of the epididymis where the absorption of fluid by epithelial cells occurs and are characterized by the presence of aquaporins (AQPs), which play a relevant role for the secretion/reabsorption dynamics of luminal fluid during sperm transport and maturation (Yeste et al. 2017). In particular, estrogens regulate the AQP1 and AQP9 activity. A reduced expression of AQP1 and AQP9 in the epididymis and efferent ducts was observed in estrogen receptor α knockout mice indicating that estrogens regulate the fluid absorption in these ducts (Carrageta et al. 2020).

Then, all along the epididymal tubule, further absorption causes a consistent increase of sperm concentration in the cauda epididymis. The concentration of spermatozoa in the epididymis reflects the sperm concentration in semen, an important parameter for male fertility (James et al. 2020).

The ionic composition of epididymal fluids is different from that observed in other fluids of the body because it contains a lower overall concentration of Na^+, Cl^-, Ca^{2+} (except for human), and HCO_3^- ions than those that have been reported in blood plasma. These ionic concentrations are, in turn, tightly associated with the regulation of luminal acidification that helps keep spermatozoa in a dormant condition.

The most relevant process occurring during epididymal transit and required for normal male fertility is sperm maturation, consisting in several changes including the acquisition of motility (Fig. 6.3).

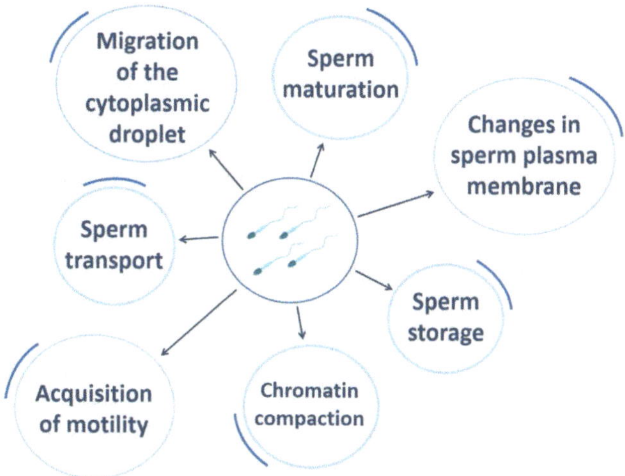

Fig. 6.3 Graphical illustration reporting the most important functions of epididymis

Each region of the epididymis is characterized by a specific luminal environment; the differences between regions are due to the cell composition of the epithelium and hormonal regulation. The progressions of the spermatozoa along the epididymis also allow changes in chromatin compaction, plasma membrane composition (Fig. 6.3), cytoskeletal structure, protein payload, and noncoding RNA payload.

Acquisition of Motility Spermatozoa released from the testis are immotile and are unable to complete any progressive movement. The maturation accounts for both a quantitative increase in the percentage of motile spermatozoa and a qualitative difference in motility pattern. In fact, testicular spermatozoa are immotile or display only a weak twitch of the flagellum. Spermatozoa released from the epididymal caput swim in a circular trajectory, while spermatozoa from the cauda epididymis show progressive motility (Robaire et al. 2006). Importantly, some signaling pathways have been associated with sperm motility, and evidence suggests that sperm may have functional flagellar machinery that is activated during epididymal transit (see Sect. 3.5).

Migration of the Cytoplasmic Droplet This phenomenon is mainly studied in animals and less known in humans for the already reported trouble in finding human epididymal tissue.

It is well known that during normal spermiogenesis, one of the most critical final events is the phagocytosis of excess spermatid cytoplasm by Sertoli cells. As a result, only a small cytoplasmic remnant, known as the "cytoplasmic droplet," remains in the elongated spermatid after its release from the germinal epithelium. The caudal migration of the cytoplasmic droplet from the neck toward the annulus, which connects the midpiece to the principal piece (see Sect. 3.3), represents a typical morphological change that spermatozoa undergo during epididymal transit (Fig. 6.3; Cooper 2011). The clear cells normally take up the contents of cytoplasmic droplets. Proteomic analyzes of purified mouse cytoplasmic droplet revealed the presence of numerous enzymes associated with energy metabolism and other proteins involved in Golgi apparatus, endoplasmic reticulum, and endosome activities. These observations can suggest a dynamic state of protein turnover in epididymal spermatozoa, even though other studies need to confirm these data (Hermo et al. 2019). In addition, cytoplasmic droplet is involved in the ion homeostasis regulation of spermatozoa. Cytoplasmic droplet contains K^+, Cl^-, and water channels, which may play a role in the regulation of sperm volume during transport through epididymal regions characterized by different luminal environment. Intracellular Ca^{2+} is also accumulated in this "organelle," responsible for maintaining low levels of tyrosine phosphorylation in immature sperm (Gervasi and Visconti 2017).

Changes in the Sperm Plasma Membrane Many potential mediators of sperm–egg interaction are acquired, or modified, in the epididymis, demonstrating the significant contribution of this part of the male reproductive system to successful fertilization. One of the most important events characterizing the epididymal sperm maturation involves the remodeling of the sperm membrane composition (Fig. 6.3), in particular, of phospholipids and sterols. The membrane of spermatozoa is particularly rich in polyunsaturated fatty acids (PUFAs) that represent a preferred target for reactive oxygen species causing lipid peroxidation. Such lipid peroxidation covers a physiological role being relevant for different phases of the fertilization process such as sperm hyperactivation, capacitation, acrosome reaction (James et al. 2020).

The modifications of plasma membrane lipids are species-specific; for example, a marked decrease in sperm cholesterol content was observed in several species including sheep, cattle, and rodents. In mice and rats, during the passage from the caput to the cauda of the epididymis, the total concentration of phospholipids in spermatozoa decreases; on the contrary, in pigs and cattle the total phospholipid amount remains the same. Irrespective to species, the cholesterol/phospholipid ratio decreases, inducing an increase of the membrane fluidity, relevant to trigger the capacitation, the acrosome reaction, and the membrane fusion during fertilization (Gautier and Aurich 2022).

The modification of the lipid pattern of the plasma membrane depends on specific lipid cross talk between spermatozoa and the epididymosomes (Barrachina et al. 2022; Roca et al. 2022), vesicles produced by epididymal epithelium that will be discussed later.

Simplifying and summarizing, as sperm pass through the caput, their membranes remain flexible to allow lipid remodeling in the corpus and cauda, where they become more stable and better able to resist damage during storage in the cauda epididymidis.

During epididymal transit, both lipids and proteins reorganize, leading to the formation of lipid rafts involved in signal transduction, cholesterol content decreases and PUFA content increases, conferring more fluidity properties to the sperm membrane.

After the spermatozoa are released from the *ductus deferens*, spatial, and temporal modifications of the sperm head membrane surface continue with re- and decoating events induced by the male accessory fluid called the *seminal plasma* (Leahy and Gadella 2011).

Chromatin Compaction During spermiogenesis, the round spermatids acquire final nuclear shape and replace somatic histones with protamines. Histones that organize the DNA into nucleosomes are gradually replaced by testis-specific histone variants, by transitions proteins Tnp1 and Tnp2 and finally by protamines. Sperm DNA-protamine interaction leads to a typical sperm chromatin conformation that includes the sperm DNA coiled into toroidal subunits, the "doughnut loops." During epididymal maturation, the nucleus is further condensed (Fig. 6.3) by active formation of disulphide bridges. The nuclear enzyme sperm glutathione peroxidase 4 (nGPx4) forms inter- and intra-protamine disulphide bonds between thiol groups of the cysteine-rich protamines (O'Flaherty 2019; Puglisi et al. 2012). In fact, sperm from nGPx4-KO mice were more prone to chromatin decondensation than those from wild-type mice at all stages of epididymal maturation, providing the conclusion that nGPx4 is required for a correct sperm chromatin compaction.

In addition, it was reported that peroxiredoxin 6 (PRDX6), an antioxidant enzyme located in the spermatozoa, also plays a role in sperm chromatin compaction. Infertile men have lower levels of sperm PRDX6 associated with low sperm motility and high DNA damage (Ozkosem et al. 2015) and knockout male mice for *Prdx6* showed higher levels of free thiols compared with wild-type controls (Ozkosem et al. 2016).

6.1.3 Gene Expression Patterns Along the Epididymis

The epididymis is an ideal organ for the study of the mechanisms of gene expression control, as it presents a highly segmented transcriptome along the duct, at least in rodent species. However, information on these mechanisms is scarce in humans and rich in rodents.

Many factors have been shown to be involved in gene expression. Steroid hormones are the best-known regulatory factors of epididymal differentiation and function. Other non-steroidal factors of testicular origin, called "lumicrine factors" can play a role in epididymal gene expression and differentiation such as neural epidermal growth factor-like-like 2 (Kiyozumi et al. 2020).

In rodent species, the different epididymal regions including the initial segment are well developed and defined, whereas in human epididymis, the zonation is not well-defined, being divided into the poorly defined caput, corpus, and cauda segments dissected according to the anatomy of the organ. The study of gene expression profile in humans revealed two regions in the human epididymis: the caput and the distal corpus/cauda region (Légaré and Sullivan 2020).

In contrast, in other species, in particular in rodents, the analyzed genes where up or down regulated several times between the different areas of epididymis (Liu et al. 2024) indicating a more marked specificity of gene expression than that observed in humans (Sullivan et al. 2019).

6.1.4 The Epididymis as a Sperm Storage

The segment involved in the sperm storage (Fig. 6.3) is the cauda (Cornwall 2009), the last part of epididymis. The luminal environment generated by the cells of epithelium preserve the spermatozoa in a quiescent state; many compounds involved in this phenomenon are mostly unknown. When the spermatozoa are mixed with seminal plasma and are ejaculated, increase their metabolism, and start a new adventure in the female reproductive tract.

6.1.5 Sperm Epididymal Transit and Protein Signature

Many studies deal with the identification of specific proteins playing a role in sperm maturation through epididymal transit. Protein composition is different in the different epididymal segments that therefore are characterized by their own transcriptome, proteome, and secretome signatures. The molecular basis of this marked regionalization is not completely clarified, although the involvement of androgens and non-steroidal testicular factors was suggested (Sipilä and Björkgren 2016).

As well as the composition of epididymal fluid, also the spermatozoa are sequentially modified during their trip along the epididymis.

Castillo et al. (2018) identified 165 human sperm proteins acquired presumably in extra-testicular environment. Spermatozoa are reputed to be unable to synthesize proteins; for this reason, the proteins appearing in them during epididymal transit are acquired and incorporated. Gervasi and Visconti (2017) suggested several mechanisms for protein transfer:

1. spermatozoa acquire proteins from epididymal fluid,
2. the protein load is performed by the tight contact of spermatozoa with epididymal epithelial cells,
3. epididymosomes, a sort of membranous vesicles like exosomes released by epididymal epithelium, prime spermatozoa with proteins, noncoding RNA, lipids (Fig. 6.2).

Thus, the spermatozoa are loaded with different molecules from the epididymis by means of diverse mechanisms that up to now remain elusive.

Epididymal maturation is not only associated with changes in protein content, but also with post-translational protein modifications. Metalloprotease-disintegrin protein family that are re-localized from the whole sperm head to the posterior head undergoes proteolytic modifications. Phosphorylation/dephosphorylation is one of the most common mechanisms of proteins changes. In general, dephosphorylation is associated with epididymal maturation, but this is not true for all proteins, for example, mitochondrial ATPase F1 and Izumo1, a protein essential for sperm-egg fusion (see Sect. 3.2.1). Within immature spermatozoa (from the proximal caput region), Izumo1 is phosphorylated in two amino acid residues, whereas mature spermatozoa from the cauda epididymis show seven sites of phosphorylation in this protein (Young et al. 2015).

Glycosylation of both N- or O-linked residues is a major post-translational modification that occurs within developing spermatozoa and is being demonstrated to be important is several key areas. Several glycoproteins are associated with the spermatozoa during epididymal descent. The vast majority of glycosyltransferase activity is associated within the lumen of epididymis and is plausibly involved in the changes of composition of sperm glycocalyx during epididymal maturation. One of the most important modifications regards the incorporation of negatively charged sialic acid responsible for a net-negative change in the external sperm surface charge that play a possible inhibitory role for capacitation (Baker 2016).

6.2 Epididymosomes

As stated before, epididymosomes are vesicles like exosomes (30–150 nm in diameter). Exosomes are produced in endosomal compartment of most eukaryotic cells, released in the extracellular environment and present in many biological fluids. They carry a different array of macromolecules, a sort of molecular signature of the producing cells, including proteins, lipids, nucleic acids (James et al. 2020; Moghassemi et al. 2024).

Epididymosomes are released from the epididymal epithelium, in particular from the principal cells and act as efficient carriers of different molecules (Fig. 6.2). At structural level, they are vesicles surrounded by a phospholipid bilayer and transmembrane proteins as CD9 and Flotillin-1 (Paul et al. 2021); the inner components include proteins, noncoding RNAs and distinct set of lipids that are transferred to spermatozoa during the transit on the different epididymal regions (Sullivan and Saez 2013). The composition of these vesicles is region-specific, thus the epididymosomes from different epididymal regions are loaded with different subclasses of macromolecules (Sullivan 2015).

Paul et al. (2021) reported the proteins identified in the epididymosomes of different species, in particular in bull, mouse, and rat. A proteomic analysis of human epididymosomes allowed the identification of 146 different proteins among them enzymes (27%), adhesion molecules (14%), carriers (13%), proteins involved in signal transduction (12%), and others as structural proteins, chaperones, and defense molecules (Thimon et al. 2008).

The ability of epididymosomes to carry small RNA cargo has been demonstrated in recent studies (Reilly et al. 2016). The in vitro treatment of spermatozoa with epididymosomes revealed an increase in small RNA amount in these cells (Sharma et al. 2018). Therefore, a role of epididymosomes in remodeling the sperm small RNA profile during epididymal transit is plausible. In fact, the epididymosomes sampled from the different segments of the epididymis are enriched in different small RNA cargos corresponding with the changes observed in spermatozoa during epididymal transit (Trigg et al. 2019).

The contact and interaction between epididymosomes and sperm crossing the epididymis (Fig. 6.2) have been proven using a variety of techniques, including the use of fluorescent markers, transgenic mice, and biotinylated proteins (Barrachina et al. 2022).

Zhou et al. (2019) studied the epididymosome-sperm interactions and discovered that the epididymosomes bind the receptors located in the post-acrosomal area of the sperm head. Then, epididymosomes transfer protein cargo to spermatozoa and this process is dependent on dynamin, a protein involved in the vesicle trafficking. The interest on epididymosomes lies also in the molecular profiling of their content to discover the proteins, lipids and RNA that characterize these vesicles (Barrachina et al. 2022). To underline the relevance of the role of extracellular vesicles present in seminal plasma in regulating sperm function, a recent research of Zhang et al. (2024) should be mentioned. Extracellular vesicles modulate Ca^{2+} concentrations by activating CatSper channels influencing human sperm motility.

6.3 Seminal Plasma: Not a Simple Vehicle for Spermatozoa

The semen is a protein rich body fluid composed by a cellular component, the spermatozoa (sometimes also other cell types as leukocytes, immature germ cells, and epithelial cells of urogenital tract), and a fluid part named seminal plasma.

Different male reproductive accessory glands contribute seminal plasma: mainly the seminal vesicles, prostate, bulbourethral and periurethral glands, and epididymis.

Each fraction of the seminal plasma is secreted in a sequential order and has a different array of chemical compounds and consequently a different function (Gupta and Kumar 2017). The combination of these fractions provides an optimal environment to transport the spermatozoa in the female reproductive system.

The volume of a typical ejaculate is around 3 mL corresponding to the 50th centile reported in the WHO guidelines for human semen analysis (2021) as given in Table 6.1.

Table 6.1 Values for semen volume reported in the last edition of WHO guidelines for semen analysis (2021)

Centiles	2.5th	5th	10th	25th	50th	75th	90th	95th	97.5th
Semen volume (mL)	1.0	1.4	1.8	2.3	3.0	4.2	5.5	6.2	6.9

Normal semen shows an opalescent, white, or gray color and its density ranges from 1.043 and 1.102 g/mL. It is characterized by the presence of high concentration of organic compounds rather than inorganic ions; this is the reason of its high osmolality.

The semen can be more transparent in case of oligozoospermia/severe oligozoospermia (low sperm concentration). The color can be different in several conditions: slightly yellowish if the abstinence time is prolonged, red-brown when red blood cells are present in the semen sample (WHO 2021). The pH of the human semen physiologically ranges from 7.2 to 7.8 and these slightly alkaline properties help to neutralize the vaginal acidic environment.

Therefore, human seminal plasma is a complex fluid that contains different constituents such as proteins, lipids, carbohydrates, hormones, inorganic ions as zinc, cell-free biopolymers as DNA, RNA, and miRNA. The exact function of all single components of the seminal plasma has not yet been discovered. Seminal plasma is not a mere vehicle capable to deliver spermatozoa to the female, but it influences fertilization, pregnancy, and embryo development. Much recent research found in the literature indicates that the study of biomarkers in seminal plasma represent a deeply investigated topic and a crucial and very stimulating area of research. Finding biomarkers in this fluid means discovering molecules marker of physiological and pathological conditions new knowledge on the diagnosis and treatment of reproductive health problems, including infertility (Álvarez-Rodríguez and Martinez-Pastor 2021; Vashisht and Gahlay 2024). At the present, we know that several semen components are markers of the performance of each reproductive accessory gland (Moustakli et al. 2024; Wang et al. 2020). For example, the seminal vesicle secretions include fructose, the main energy source of sperm in semen, semenogelin, a protein involved in the formation of the coagulum in the semen, prostaglandins, phospholipids, fibronectin, lactoferrin, protein C inhibitor, and prolactin-inducible protein (Drabovich et al. 2014; Rodríguez-Martínez et al. 2011). The assessment of prostate-specific antigen that hydrolyzes semenogelin allowing the liquefaction of the coagulum, prostatic acid phosphatase, cysteine-rich prostate-specific protein-94, citric acid, and zinc, particularly concentrated in seminal plasma, are markers of the prostatic gland function. Epididymal function is monitored by assessing glycerophosphocholine, free-L-carnitine, and neutral α-glucosidase (Gupta and Kumar 2017). Finally, the bulbourethral glands produce sialic acid, mucin, and galactose, all compounds able to improve the efficacy of sperm transfer and permanence in female reproductive tract.

However, many other markers are studied, and particular interest is raised on those related to male infertility (Druart et al. 2019; Fietz et al. 2024; Kumar and Singh 2020; Parvin et al. 2024).

For example, the epididymal protein extracellular matrix protein 1 seems to be a reliable marker in differentiating obstructive azoospermia from non-obstructive azoospermia (Ješeta et al. 2023). Sutovsky et al. (2004) demonstrated that ubiquitin is a powerful biomarker that can assess male fertility and human semen quality, since an inverse relationship between ubiquitin levels and sperm motility was observed.

Many proteins that potentially can have a crucial role in male infertility are worth exploring. One interesting protein is TEX101, which during epididymal maturation is cleaved from the surface of spermatozoa, and it is released on the seminal plasma (Fujihara et al. 2013; Schiza et al. 2014). The cleavage is essential for sperm maturation; for this reason, the levels of TEX101 detected in seminal plasma indicate the efficiency of spermatogenesis (high: normal spermatogenesis; low: abnormal spermatogenesis) and can be used as a marker of male infertility (Drabovich et al. 2013).

Lipocalin-type prostaglandin D synthase is another protein enveloping the spermatozoa and released in the seminal plasma. This protein acts as a carrier of retinoid, essential fatty acids, and other compounds needed to support the germ cell development and maturation and its differential levels in seminal plasma can indicate different spermatogenesis conditions and, in particular, the non-obstructive azoospermia, avoiding the testicular biopsy (Heshmat et al. 2008; Vashisht and Gahlay 2024). In addition, the protein macrophage migration inhibitory factor (MIF) plays a relevant role in the redox homeostasis of thiol groups. Alteration of physiological levels of MIF was observed in seminal plasma of patients with altered sperm parameters, indicating the potential involvement of MIF as a biomarker in such cases of male infertility (Vashisht and Gahlay 2024).

Nowicka-Bauer et al. (2022) focused their attention on Heat shock-related 70 kDa protein 2 (HSPA2), a chaperon protein linked to spermatogenesis status. The HSPA2 dosed in the seminal plasma represents a potential marker for spermatogenesis efficiency, since azoospermic men are devoid of this seminal plasma protein. In addition, they observed that HSPA2 might predict the outcome of assisted reproductive techniques (ART).

In addition, several molecules related to oxidative stress and antioxidants have been indicated as potential biomarker for fertility/infertility (Moretti et al. 2022; Moustakli et al. 2024).

Several other biochemical components are associated with human semen pathological conditions. For example, albumin, total protein levels, and the γ-glutamyl transferase activity are increased in seminal plasma of asthenozoospermic patients. High level of seminal ferritin was associated with infertility and with the presence of immature germ cells in human semen. In addition, smoking habit was linked with decreased levels of folic acid and estradiol (Collodel et al. 2019). At this purpose, also the seminal plasma composition of infertile patients with varicocele, leukocytospermia, and idiopathic infertility differs in several biochemical parameters. For example, F_2-isoprostanes, derived from the non-enzymatic peroxidation of arachidonic acid and marker of oxidative stress, were increased in seminal plasma of patients with varicocele characterized by sperm immaturity, whereas seminal iron and ferritin levels are increased in cases with leukocytospermia and sperm necrosis (Collodel et al. 2021). These data indicate that seminal plasma composition deserve further studies and attention because the different amount of several compounds seem to be associated with specific reproductive pathologies. Recently, it has been found that seminal Resolvin D1 (RvD1, see Sect. 4.1.1), a specialized pro-resolving mediator produced by docosahexaenoic acid exerting anti-inflammatory and pro-resolving activities, increased along with other seminal markers of oxidative stress and inflammation as fatty acids content and clinical biomarkers (Signorini et al. 2022).

Another aspect considers the presence of a significant number of extracellular vesicles (Parvin et al. 2024), mainly prostasomes, in seminal plasma. Saez and Sullivan (2016) identified more than 1280 proteins in prostasomes playing a significant role in sperm motility and capacitation. Differences in the amount and expression of proteins carried by prostasomes were observed between normozoospermic men and infertile patients (Panner Selvam et al. 2019).

In addition, miRNAs assessed in the seminal plasma have also emerged as potential hallmarks of spermatogenesis status, although additional investigations are necessary to determine their diagnostic utility (Vashisht and Gahlay 2020).

Other than sperm quality and reproductive pathologies, the concentration of several seminal plasma molecules or combination of molecules make this fluid a good potential source for possible biomarkers that could predict ART outcomes. Since at present semen quality parameters cannot accurately predict ART outcomes, future studies considering the seminal plasma are needed and urgent (van den Berg et al. 2024). Recently, Płaczkowska et al. (2024) found a connection between several blood serum and seminal plasma parameters (low-density lipoprotein, AMP-activated protein kinase, and carnitine), thus suggesting a systemic influence on seminal plasma composition. These observations indicate the intricate interplay between the systemic conditions of an individual, its seminal plasma composition and sperm parameters opening future directions in the male infertility research. Probably, the right way to approach this problem includes an integrated methodology considering the field of genomic, epigenomic, transcriptomic, proteomic, and metabolomic changes in the seminal plasma of infertile men for identifying new biomarkers (Llavanera et al. 2022). Obviously, more investigations and advancements are necessary before these biomarkers can reach their full diagnostic potential; for this reason, this is an open and stimulating area of research that need to be developed in the next few years.

6.4 Conclusion

The main problem in the study of human epididymis is the trouble in retrieving the material; for this reason, the research on animal models is of pivotal importance for the knowledge of physiology of this structure. The most interesting issues to be explored in humans are the mechanisms of interaction between epididymal epithelium and spermatozoa, the regulation of gene expression in the different segments of epididymis and, in particular, the role of epididymosomes in priming spermatozoa with different molecules. Since epididymosomes influence epigenetics of human spermatozoa, they can have an impact on the outcome of assisted fertilization when the spermatozoa are recovered from different sites along the male tract.

Because the conventional sperm parameters, evaluated during semen analysis, are not predictive of true fertility potential, a relevant source of new biomarkers for infertility or assisted fertilization outcome is the seminal plasma. This fluid, produced by male reproductive glands, regulates several processes as capacitation, protection of spermatozoa and even interaction with the zona pellucida of oocyte. In addition, the quality of seminal plasma can influence the results of ART and semen cryopreservation procedures that are a routine process of living sperm preservation at low temperatures for long period, used as side technique in ART.

Therefore, seminal plasma is a sample that is obtained with non-invasive method and shows the potential in discovering markers related to reproductive disorders.

At the same time, despite the scarcity of material available, the study of human epididymis is crucial to implement the knowledge on sperm physiology in normal and altered conditions and the real impact of epididymal sperm maturation on the ART outcome.

The Spermatozoa in the Cold: Suitable Cells for Cryopreservation

Compared to other cell types, spermatozoa seem to be less sensitive to damages caused by the freezing process, essentially for their particular membrane lipid composition and for low water content due to the virtual lack of cytoplasm. Sperm cryopreservation is one of the most effective techniques in preserving fertility. A wide range of males can benefit from sperm freezing, among them:

- oncologic patients undergoing chemotherapy or radiotherapy, both procedures affecting spermatogenesis;
- men undergoing vasectomy or with testicular trauma;
- healthy donors, whose spermatozoa can be frozen for sperm banking, a common global procedure necessary for heterologous fertilization protocol;
- patients with severe oligozoospermia, or azoospermia, who have undergone testicular sperm extraction, avoiding repeated surgical procedures.

Although spermatozoa are suitable for freezing, this procedure affects sperm function damaging motility, vitality, membrane and acrosomal integrity, mitochondrial structure and activity (Fig. 6.4) and induces DNA fragmentation, alterations of mRNA expression, and epigenetic status, changes in metabolic and protein profiles. The observed post-thaw reduced sperm quality can be due to different mechanisms such as rapid temperature fluctuations, ice crystal formation, osmotic stress, and increased production of reactive oxygen species that causes oxidative insults (Fig. 6.4). Intervention to improve the values of sperm parameters after thawing are advisable.

Several strategies to minimize freezing-thawing sperm damage include the use of different cryoprotectant agents, antifreeze proteins, glycoproteins, and supplementation of media with different antioxidants. In contrast, some research proposed moderate "offensive strategies" applying a mild stress before freezing, which causes a reaction and adaptation of spermatozoa. In other words, this is a sort of training that spermatozoa undertake before the freezing process. The research in this field is appealing because of the urgent need to improve the cryopreservation procedures and consequently, fertilization results and pregnancy rates.

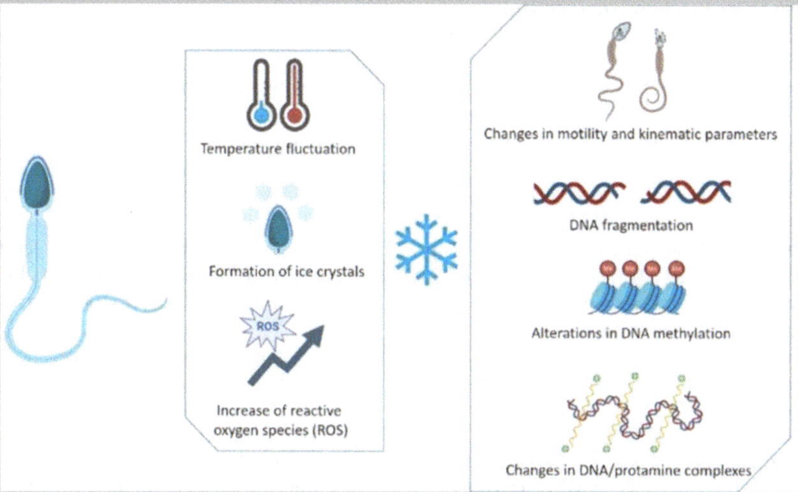

Fig. 6.4 Diagram showing several conditions that during the freezing-thawing process can damage spermatozoa. Temperature fluctuation, formation of ice crystals, and increase of reactive oxygen species (ROS) can cause a decrease in sperm motility and vitality, DNA, and chromatin alterations

References

Ali W, Deng K, Bian Y, Liu Z, Zou H (2023) Spectacular role of epididymis and bio-active cargo of nano-scale exosome in sperm maturation: a review. Biomed Pharmacother 164:114889. https://doi.org/10.1016/j.biopha.2023.114889

Álvarez-Rodríguez M, Martinez-Pastor F (2021) Molecular determinants of seminal plasma on sperm biology and fertility. Int J Mol Sci 22(7):3555. https://doi.org/10.3390/ijms22073555

Baker MA (2016) Proteomics of post-translational modifications of mammalian spermatozoa. Cell Tissue Res 363(1):279–287. https://doi.org/10.1007/s00441-015-2249-x

Barrachina F, Battistone MA, Castillo J, Mallofré C, Jodar M, Breton S, Oliva R (2022) Sperm acquire epididymis-derived proteins through epididymosomes. Hum Reprod 37(4):651–668. https://doi.org/10.1093/humrep/deac015

Breton S, Ruan YC, Park YJ, Kim B (2016) Regulation of epithelial function, differentiation, and remodeling in the epididymis. Asian J Androl 18(1):3–9. https://doi.org/10.4103/1008-682X.165946

Carrageta DF, Bernardino RL, Soveral G, Calamita G, Alves MG, Oliveira PF (2020) Aquaporins and male (in)fertility: expression and role throughout the male reproductive tract. Arch Biochem Biophys 679:108222. https://doi.org/10.1016/j.abb.2019.108222

Castillo J, Jodar M, Oliva R (2018) The contribution of human sperm proteins to the development and epigenome of the preimplantation embryo. Hum Reprod Update 24:535–555

Collodel G, Nerucci F, Signorini C, Iacoponi F, Moretti E (2019) Associations between biochemical components of human semen with seminal conditions. Syst Biol Reprod Med 65(2):155–163. https://doi.org/10.1080/19396368.2018.1548668

Collodel G, Signorini C, Nerucci F, Gambera L, Iacoponi F, Moretti E (2021) Semen biochemical components in varicocele, Leukocytospermia, and idiopathic infertility. Reprod Sci 28(1):91–101. https://doi.org/10.1007/s43032-020-00260-0

Cooper TG (2011) The epididymis, cytoplasmic droplets and male fertility. Asian J Androl 13(1):130–138. https://doi.org/10.1038/aja.2010.97

Cornwall GA (2009) New insights into epididymal biology and function. Hum Reprod Update 15:213–227. https://doi.org/10.1093/humupd/dmn055

Cyr DG, Gregory M, Dubé E, Dufresne J, Chan PT, Hermo L (2007) Orchestration of occludins, claudins, catenins and cadherins as players involved in maintenance of the blood-epididymal barrier in animals and humans. Asian J Androl 9(4):463–475. https://doi.org/10.1111/j.1745-7262.2007.00308.x

Drabovich AP, Dimitromanolakis A, Saraon P, Soosaipillai A, Batruch I, Mullen B, Jarvi K, Diamandis EP (2013) Differential diagnosis of azoospermia with proteomic biomarkers ECM1 and TEX101 quantified in seminal plasma. Sci Transl Med 5(212):212ra160. https://doi.org/10.1126/scitranslmed.3006260

Drabovich AP, Saraon P, Jarvi K, Diamandis EP (2014) Seminal plasma as a diagnostic fluid for male reproductive system disorders. Nat Rev Urol 11(5):278–288. https://doi.org/10.1038/nrurol.2014.74

Druart X, Rickard JP, Tsikis G, de Graaf SP (2019) Seminal plasma proteins as markers of sperm fertility. Theriogenology 137:30–35. https://doi.org/10.1016/j.theriogenology.2019.05.034

Dufresne J, Gregory M, Pinel L, Cyr DG (2024) Three-dimensional cell culture of epididymal basal cells and organoids: a novel tool for toxicology. Curr Protoc 4(1):e975. https://doi.org/10.1002/cpz1.975

Elfgen V, Mietens A, Mewe M, Hau T, Middendorff R (2018) Contractility of the epididymal duct: function, regulation and potential drug effects. Reproduction 156(4):R125–R141. https://doi.org/10.1530/REP-17-0754

Fietz D, Sgaier R, O'Donnell L, Stanton PG, Dagley LF, Webb AI, Schuppe HC, Diemer T, Pilatz A (2024) Proteomic biomarkers in seminal plasma as predictors of reproductive potential in azoospermic men. Front Endocrinol (Lausanne) 15:1327800. https://doi.org/10.3389/fendo.2024.1327800

Fujihara Y, Tokuhiro K, Muro Y, Kondoh G, Araki Y, Ikawa M, Okabe M (2013) Expression of TEX101, regulated by ACE, is essential for the production of fertile mouse spermatozoa. Proc Natl Acad Sci USA 110(20):8111–8116. https://doi.org/10.1073/pnas.1222166110

Gautier C, Aurich C (2022) Fine feathers make fine birds. The mammalian sperm plasma membrane lipid composition and effects on assisted reproduction. Anim Reprod Sci 246:106884. https://doi.org/10.1016/j.anireprosci.2021.106884

Gervasi MG, Visconti PE (2017) Molecular changes and signaling events occurring in spermatozoa during epididymal maturation. Andrology 5(2):204–218. https://doi.org/10.1111/andr.12320

Gupta S, Kumar A (2017) The human semen. https://doi.org/10.1007/978-981-10-3695-8_11

Hermo L, Oliveira RL, Smith CE, Au CE, Bergeron JJM (2019) Dark side of the epididymis: tails of sperm maturation. Andrology 7(5):566–580. https://doi.org/10.1111/andr.12641

Heshmat SM, Mullen JB, Jarvi KA, Soosaipillai A, Diamandis EP, Hamilton RJ, Lo KC (2008) Seminal plasma lipocalin-type prostaglandin D synthase: a potential new marker for the diagnosis of obstructive azoospermia. J Urol 179(3):1077–1080. https://doi.org/10.1016/j.juro.2007.10.070

James ER, Carrell DT, Aston KI, Jenkins TG, Yeste M, Salas-Huetos A (2020) The role of the epididymis and the contribution of epididymosomes to mammalian reproduction. Int J Mol Sci 21(15):5377. https://doi.org/10.3390/ijms21155377

Ješeta M, Pospíšilová A, Mekiňová L, Franzová K, Ventruba P, Lousová E, Kempisty B, Oždian T, Žáková J, Crha I (2023) Non-invasive diagnostics of male spermatogenesis from seminal plasma: seminal proteins. Diagnostics (Basel) 13(15):2468. https://doi.org/10.3390/diagnostics13152468

Kiyozumi D, Noda T, Yamaguchi R, Tobita T, Matsumura T, Shimada K, Kodani M, Kohda T, Fujihara Y, Ozawa M, Yu Z, Miklossy G, Bohren KM, Horie M, Okabe M, Matzuk MM, Ikawa M (2020) NELL2-mediated lumicrine signaling through OVCH2 is required for male fertility. Science 368(6495):1132–1135. https://doi.org/10.1126/science.aay5134

Kumar N, Singh NK (2020) Emerging role of novel seminal plasma bio-markers in male infertility: a review. Eur J Obstet Gynecol Reprod Biol 253:170–179. https://doi.org/10.1016/j.ejogrb.2020.08.015

Leahy T, Gadella BM (2011) Sperm surface changes and physiological consequences induced by sperm handling and storage. Reproduction 142(6):759–778. https://doi.org/10.1530/REP-11-0310

Légaré C, Sullivan R (2020) Differential gene expression profiles of human efferent ducts and proximal epididymis. Andrology 8(3):625–636. https://doi.org/10.1111/andr.12745

Liu MM, Feng XL, Qi C, Zhang SE, Zhang GL (2024) The significance of single-cell transcriptome analysis in epididymis research. Front Cell Dev Biol 12:1357370. https://doi.org/10.3389/fcell.2024.1357370

Llavanera M, Delgado-Bermúdez A, Ribas-Maynou J, Salas-Huetos A, Yeste M (2022) A systematic review identifying fertility biomarkers in semen: a clinical approach through omics to diagnose male infertility. Fertil Steril 118(2):291–313. https://doi.org/10.1016/j.fertnstert.2022.04.028

Moghassemi S, Dadashzadeh A, Sousa MJ, Vlieghe H, Yang J, León-Félix CM, Amorim CA (2024) Extracellular vesicles in nanomedicine and regenerative medicine: a review over the last decade. Bioactive Materials 36:126–156. https://doi.org/10.1016/j.bioactmat.2024.02.021

Moretti E, Signorini C, Ferretti F, Noto D, Collodel G (2022) A study to validate the relevance of semen F_2-Isoprostanes on human male infertility. Int J Environ Res Public Health 19(3):1642. https://doi.org/10.3390/ijerph19031642

Moustakli E, Zikopoulos A, Skentou C, Stavros S, Sofikitis N, Georgiou I, Zachariou A (2024) Integrative assessment of seminal plasma biomarkers: a narrative review bridging the gap between infertility research and clinical practice. J Clin Med 13(11):3147. https://doi.org/10.3390/jcm13113147

Nowicka-Bauer K, Malcher A, Włoczkowska O, Kamieniczna M, Olszewska M, Kurpisz MK (2022) Evaluation of seminal plasma HSPA2 protein as a biomarker of human spermatogenesis status. Reprod Biol 22(1):100597. https://doi.org/10.1016/j.repbio.2021.100597

O'Flaherty C (2019) Orchestrating the antioxidant defenses in the epididymis. Andrology 7(5):662–668. https://doi.org/10.1111/andr.12630

Ozkosem B, Feinstein SI, Fisher AB, O'Flaherty C (2015) Advancing age increases sperm chromatin damage and impairs fertility in peroxiredoxin 6 null mice. Redox Biol 5:15–23. https://doi.org/10.1016/j.redox.2015.02.004

Ozkosem B, Feinstein SI, Fisher AB, O'Flaherty C (2016) Absence of peroxiredoxin 6 amplifies the effect of oxidant stress on mobility and SCSA/CMA3 defined chromatin quality and impairs fertilizing ability of mouse spermatozoa. Biol Reprod 94(3):68. https://doi.org/10.1095/biolreprod.115.137646

Panner Selvam MK, Agarwal A, Pushparaj PN, Baskaran S, Bendou H (2019) Sperm proteome analysis and identification of fertility-associated biomarkers in unexplained male infertility. Genes (Basel) 10(7):522. https://doi.org/10.3390/genes10070522

Parvin A, Erabi G, Alemi A, Rezanezhad A, Maleksabet A, Sadeghpour S, Taheri-Anganeh M, Ghasemnejad-Berenji H (2024) Seminal plasma proteomics as putative biomarkers for male infertility diagnosis. Clin Chim Acta 561:119757. https://doi.org/10.1016/j.cca.2024.119757

Paul N, Talluri TR, Nag P, Kumaresan A (2021) Epididymosomes: a potential male fertility influencer. Andrologia 53(9):e14155. https://doi.org/10.1111/and.14155

Płaczkowska S, Rodak K, Kmieciak A, Gilowska I, Kratz EM (2024) Exploring correlations: human seminal plasma and blood serum biochemistry in relation to semen quality. PLoS One 19(6):e0305861. https://doi.org/10.1371/journal.pone.0305861
Puglisi R, Maccari I, Pipolo S, Conrad M, Mangia F, Boitani C (2012) The nuclear form of glutathione peroxidase 4 is associated with sperm nuclear matrix and is required for proper paternal chromatin decondensation at fertilization. J Cell Physiol 227(4):1420–1427. https://doi.org/10.1002/jcp.22857
Reilly JN, McLaughlin EA, Stanger SJ, Anderson AL, Hutcheon K, Church K, Mihalas BP, Tyagi S, Holt JE, Eamens AL, Nixon B (2016) Characterisation of mouse epididymosomes reveals a complex profile of microRNAs and a potential mechanism for modification of the sperm epigenome. Sci Rep 6:31794. https://doi.org/10.1038/srep31794
Robaire B, Hinton BT, Orgebin-Crist MC (2006) The epididymis. In: Neill JD (ed) Knobil and Neill's physiology of reproduction, 3rd edn. Academic Press, pp 1071–1148
Roca J, Rodriguez-Martinez H, Padilla L, Lucas X, Barranco I (2022) Extracellular vesicles in seminal fluid and effects on male reproduction. An overview in farm animals and pets. Anim Reprod Sci 246:106853. https://doi.org/10.1016/j.anireprosci.2021.106853
Rodríguez-Martínez H, Kvist U, Ernerudh J, Sanz L, Calvete JJ (2011) Seminal plasma proteins: what role do they play? Am J Reprod Immunol 66(Suppl 1):11–22. https://doi.org/10.1111/j.1600-0897.2011.01033.x
Saez F, Sullivan R (2016) Prostasomes, post-testicular sperm maturation and fertility. Front Biosci (Landmark Ed) 21(7):1464–1473. https://doi.org/10.2741/4466
Schiza CG, Jarv K, Diamandis EP, Drabovich AP (2014) An emerging role of TEX101 protein as a male infertility biomarker. EJIFCC 25(1):9–26
Sharma U, Sun F, Conine CC, Reichholf B, Kukreja S, Herzog VA, Ameres SL, Rando OJ (2018) Small RNAs are trafficked from the epididymis to developing mammalian sperm. Dev Cell 46:481–494.e6. https://doi.org/10.1016/j.devcel.2018.06.023
Signorini C, Moretti E, Noto D, Micheli L, Ponchia R, Collodel G (2022) Fatty acid oxidation and pro-resolving lipid mediators are related to male infertility. Antioxidants (Basel) 11(1):107. https://doi.org/10.3390/antiox11010107
Sipilä P, Björkgren I (2016) Segment-specific regulation of epididymal gene expression. Reproduction 152(3):R91–R99. https://doi.org/10.1530/REP-15-0533
Sullivan R (2015) Epididymosomes: a heterogeneous population of microvesicles with multiple functions in sperm maturation and storage. Asian J Androl 17:726–729
Sullivan R, Belleannee C (2018) Cell biology of the epididymis. In: Jegou B, Skinner M (eds) Encyclopedia of reproduction, 2nd edn. Elsevier, New York, NY, USA, pp 286–291
Sullivan R, Mieusset R (2016) The human epididymis: its function in sperm maturation. Hum Reprod Update 22(5):574–587. https://doi.org/10.1093/humupd/dmw015
Sullivan R, Saez F (2013) Epididymosomes, prostasomes, and liposomes: their roles in mammalian male reproductive physiology. Reproduction 146(1):R21–R35. https://doi.org/10.1530/REP-13-0058
Sullivan R, Légaré C, Lamontagne-Proulx J, Breton S, Soulet D (2019) Revisiting structure/functions of the human epididymis. Andrology 7(5):748–757. https://doi.org/10.1111/andr.12633
Sutovsky P, Hauser R, Sutovsky M (2004) Increased levels of sperm ubiquitin correlate with semen quality in men from an andrology laboratory clinic population. Hum Reprod 19(3):628–638. https://doi.org/10.1093/humrep/deh131
Thimon V, Frenette G, Saez F, Thabet M, Sullivan R (2008) Protein composition of human epididymosomes collected during surgical vasectomy reversal: a proteomic and genomic approach. Hum Reprod 23(8):1698–1707. https://doi.org/10.1093/humrep/den181
Trigg NA, Eamens AL, Nixon B (2019) The contribution of epididymosomes to the sperm small RNA profile. Reproduction 157(6):R209–R223. https://doi.org/10.1530/REP-18-0480
van den Berg JS, Molina NM, Altmäe S, Arends B, Steba GS (2024) A systematic review identifying seminal plasma biomarkers and their predictive ability on IVF and ICSI outcomes. Reprod Biomed Online 48(2):103622. https://doi.org/10.1016/j.rbmo.2023.103622

Varner DD (2015) Odyssey of the spermatozoon. Asian J Androl 17(4):522–528. https://doi.org/10.4103/1008-682X.153544

Vashisht A, Gahlay GK (2020) Using miRNAs as diagnostic biomarkers for male infertility: opportunities and challenges. Mol Hum Reprod 26(4):199–214. https://doi.org/10.1093/molehr/gaaa016

Vashisht A, Gahlay GK (2024) Understanding seminal plasma in male infertility: emerging markers and their implications. Andrology 12(5):1058–1077. https://doi.org/10.1111/andr.13563

Wang F, Yang W, Ouyang S, Yuan S (2020) The vehicle determines the destination: the significance of seminal plasma factors for male fertility. Int J Mol Sci 21(22):8499. https://doi.org/10.3390/ijms21228499

World Health Organization (ed) (2021) WHO laboratory manual for the examination and processing of human semen, 6th edn. World Health Organization

Yeste M, Morató R, Rodríguez-Gil JE, Bonet S, Prieto-Martínez N (2017) Aquaporins in the male reproductive tract and sperm: functional implications and cryobiology. Reprod Domest Anim 52(Suppl 4):12–27. https://doi.org/10.1111/rda.13082

Young SA, Aitken J, Baker M (2015) Phosphorylation of Izumo1 and its role in male infertility. Asian J Androl 17(5):708–710. https://doi.org/10.4103/1008-682X.156119

Zhang X, Liang M, Song D, Huang R, Chen C, Liu X, Chen H, Wang Q, Sun X, Song J, Zhang J, Kang H, Zeng X (2024) Both protein and non-protein components in extracellular vesicles of human seminal plasma improve human sperm function via CatSper-mediated calcium signaling. Hum Reprod 39(4):658–673. https://doi.org/10.1093/humrep/deae018

Zhou W, Stanger SJ, Anderson AL, Bernstein IR, De Iuliis GN, McCluskey A, McLaughlin EA, Dun MD, Nixon B (2019) Mechanisms of tethering and cargo transfer during epididymosome-sperm interactions. BMC Biol 17(1):35. https://doi.org/10.1186/s12915-019-0653-5

The Human Spermatozoon: Conclusion and Future Directions

7

Core Messages

The study of human spermatozoon is far from being exhausted and further research is mandatory with the aim to improve the assisted fertilization outcomes. Different aspects require investigations and in-depth analysis. The impact of sperm DNA fragmentation, epigenetics, and centrosome on embryo development is worth exploring. In addition to spermatozoa, a special mention should be made to seminal plasma components that have potential diagnostic significance but are seldom evaluated in clinical practice. In recent decades, the study of the impact of oxidative stress on spermatozoa and its connection to male infertility is undoubtedly one of the most advanced scientific fields. Finally, the spermatozoon is a protagonist also in the field of artificial intelligence whose algorithms are promising in the improvement of semen analysis and in addressing the challenges of sperm selection during intracytoplasmic sperm injection. In this brief excursus, the need for additional research into the structural, biochemical, genetic characteristics of human spermatozoa is presented.

Abbreviations

ART Assisted reproduction technologies
ICSI Intracytoplasmic sperm injection
ROS Reactive oxygen species

From Leeuwenhoek's discovery of spermatozoa up to the present, a significant number of scientific papers on this topic have been published. This confirms that the research on spermatozoa is appealing and has reached important milestones from

E. Moretti, G. Collodel, *The Human Spermatozoon*,
https://doi.org/10.1007/978-3-031-97868-5_7

the evolutionary and genetic point of view and in the field of assisted reproduction technologies (ART). At the same time, we believe that additional research on human and animal spermatozoa and reproductive system is mandatory for different reasons.

Particularly in the ART field, the human spermatozoon seems to have lost part of its importance. With the advent of intracytoplasmic sperm injection (ICSI), the operator picks up a spermatozoon and injects it into the oocyte, often with limited knowledge of sperm anatomy and its influence on sperm function (Mortimer 2018).

However, the success rates have not yet been fully optimized, and it is crucial to further understand the role of sperm evaluation in predicting ART outcomes. Additionally, the precise functions of different sperm structures in embryo activation and the early stages of its development remain to be elucidated.

It is evident that the standard semen analysis, the first line examination in the evaluation of male infertility, shows many limitations in predicting the chance of both natural and assisted fertilization (Schlegel 2023). The single semen parameters provide a partial indication of the fertilizing potential, and semen characteristics are not a direct expression of fertility. For these reasons, the current method of semen testing and the clinical evaluation of male factors do not provide valuable markers of reproductive competence and poorly predict ART outcome.

This remains an open question for future research.

The sperm DNA fragmentation parameter is perhaps the most useful in predicting the chance of fertilization and pregnancy (Vallet-Buisan et al. 2023), but it is not included in the initial semen analysis because the abnormality of this parameter does not, as of yet, change the initial management of an infertile couple. In addition, oocytes, in particular those of younger women, seem to repair sperm DNA if the damage is not extended (Marinaro 2023). Oocytes of women of advanced age, a frequent finding in ART setting, show limited ability to repair the sperm DNA fragmentation, a parameter that became clinically important.

Without doubt, a structure of the human spermatozoon that is worth exploring is the centrosome, located in the connecting piece next to the basal plate. In human spermatozoa a canonical proximal centriole perpendicular to the sperm axis is located at the base of the head; during maturation, the distal centriole undergoes a process named centrosome reduction (Manandhar et al. 2000) or centrosome remodeling (Avidor-Reiss et al. 2020). Following the classical hypothesis of centrosome reduction, the mammalian spermatozoa lack a distal centriole; recently this sort of dogma was revised by the "Centriole Remodeling Hypothesis" supporting the presence of an atypical highly specialized, yet functional distal centriole in spermatozoa of most mammals and insect (Fishman et al. 2018) aligned with the flagellum and origin of the sperm axoneme. Human oocytes are acentriolar and the spermatozoon contributes the centrosome to initiate pronuclear migration, syngamy, and chromosome segregation.

A structure practically neglected in human sperm is the centriole adjunct, a bunch of microtubule sorting from the proximal centriole that undergoes dynamic changes during maturation and degenerate in mature human spermatozoa. The function of the centriolar adjunct is not clear yet and the few studies available in humans suggest that the presence of this structure in ejaculated spermatozoa can be considered a sign of incomplete maturation (Garanina et al. 2019; Moretti et al. 2021). The critical functions in early human embryo depend on the integrity of sperm centrosome, which is definitely an understudied structure in human spermatozoa, in particular in altered spermatozoa often used during ART procedures.

Another developing field of research concerns sperm epigenetics such as DNA methylation, histone modifications, and non-coding RNAs. The discoveries in such a vast area of research can disclose key factors in identifying spermatogenetic defects responsible for abnormal sperm function and paternal factors contributing to embryo development (Ashapkin et al. 2023; Lismer and Kimmins 2023). In particular, miRNAs, expressed in a cell- or stage-specific manner during spermatogenesis, can represent a potential marker in diagnosing male infertility, predicting the varicocele grade, the surgical results, ART outcomes, and sperm retrieval rates in case of non-obstructive azoospermia (Shi et al. 2024).

Recently, attention has shifted not only to spermatozoa but also to seminal plasma, which, despite being rich in components with potential diagnostic value, is seldom assessed in clinical practice.

However, this field of research deserves further development in the next years and likely holds promising results in the diagnosis of male infertility and in predicting the fertilization outcome. Obviously, several pitfalls to be overcome regard the issue of standardization, the interpretation of seminal plasma markers in clinical practice, and others (Vashisht and Gahlay 2024).

A special mention should be devoted to extracellular vesicles present in the seminal plasma, rich in bioactive molecules. Even though this is a very hot topic in this study area, none of the molecules among those embedded in extracellular vesicles are recognized as biomarkers of male infertility, and candidate molecules are worth exploring (Parra et al. 2023).

Certainly, one of the most developed research areas in the last decades regards the effect of oxidative stress on spermatozoa and its role on male infertility (Dutta et al. 2019). In fact, increased levels of reactive oxygen species (ROS) and oxidative stress are common findings in many reproductive pathologies and in a part of undiagnosed infertility. When ROS generation exceeds the threshold of antioxidant clearance, or when antioxidant concentration is reduced, oxidative stress occurs damaging cell molecules such as carbohydrates, nucleic acids, proteins, and lipids (Drevet et al. 2022). Spermatozoa in particular are exposed to the oxidative damage due to the high concentration of polyunsaturated fatty acids in their plasma membrane. Numerous studies have confirmed the harmful impact of oxidative stress on semen parameters such as sperm viability, motility, and fertilizing potential as evidenced by higher levels of ROS in the semen of infertile men compared to fertile controls (Mannucci et al. 2022). The determination of oxidative stress is currently not indicated in the routine evaluation of infertile men and remains a challenging condition to be diagnosed and managed (Agarwal and Majzoub 2017).

To emphasize the dimension of this problem, recently a new category of infertility was proposed: "Male Oxidative Stress Infertility" including infertile men with oxidative stress (Alfaro Gómez et al. 2023). The lack of reliable biomarkers to quantify oxidative injury and determine the efficacy of possible treatments to decrease damages has been a crucial problem in the field of male infertility and, in general, in human physiology and pathology (Milne 2017).

In any case, comprehensive knowledge on sperm structure, maturation, motility, and fertilization ability is of pivotal importance for the development of potential therapy to counteract male infertility and the consequent decline of birth rates.

On the other hand, a comprehensive understanding of the spermatogenetic process and spermatozoa forms the foundation for one of the key ambitions of both the scientific community and society at large: the development of safe contraceptives for men, which are essential for reproductive control.

Furthermore, one cannot overlook the role of artificial intelligence algorithms and machine learning, which have recently been integrated into the field of reproduction. These technologies offer promising applications, particularly in challenging tasks such as sperm selection in ART and the objective evaluation of human semen (Cherouveim et al. 2023). For these reasons, andrologists and embryologists may greatly benefit from using these algorithms that theoretically could play a positive impact on the ART success and the analysis of human semen. Furthermore, these tools could improve over time as larger and more reliable datasets become available for training.

Finally, and importantly, spermatozoa serve as a reliable model for in vitro studies for various reasons. First, they are easy to collect and are differentiated cells endowed with motility, which is an exclusive parameter deeply influenced by the external environment. In addition, human spermatozoa do not possess DNA-repairing activity, and they virtually lack intracellular antioxidant protection, having limited ability to neutralize oxidative damage. All these characteristics make spermatozoa a reliable model to test in vitro many substances at different times and at different concentrations. The study of the effects on spermatozoa treated with different compounds should include the evaluation of several endpoints, as status of the acrosome, chromatin, motility, vitality, in order to analyze how the different sperm structures react to the treatment (Moretti et al. 2023).

Moreover, these studies may discover better conditions during the freezing/thawing procedure that is routinely applied in ART laboratories. Human sperm show a specific cryophysical behavior, and different sperm freezing protocols have been developed to avoid damage to the sperm cells (Noto et al. 2021).

At the conclusion of this book, we aim to have provided a comprehensive overview of human spermatozoa, emphasizing the need for further investigation into their genetic and biochemical aspects. We also hope to have highlighted the potential of this fascinating cell type in in vitro studies. Finally, we trust that readers will find as much enjoyment in reading this book as we have in writing it.

References

Agarwal A, Majzoub A (2017) Laboratory tests for oxidative stress. Indian J Urol 33(3):199–206. https://doi.org/10.4103/iju.IJU_9_17

Alfaro Gómez M, Fernández-Santos MDR, Jurado-Campos A, Soria-Meneses PJ, Montoro Angulo V, Soler AJ, Garde JJ, Rodríguez-Robledo V (2023) On males, antioxidants and infertility (MOXI): certitudes, uncertainties and trends. Antioxidants (Basel) 12(8):1626. https://doi.org/10.3390/antiox12081626

Ashapkin V, Suvorov A, Pilsner JR, Krawetz SA, Sergeyev O (2023) Age-associated epigenetic changes in mammalian sperm: implications for offspring health and development. Hum Reprod Update 29(1):24–44. https://doi.org/10.1093/humupd/dmac033

Avidor-Reiss T, Carr A, Fishman EL (2020) The sperm centrioles. Mol Cell Endocrinol 518:110987. https://doi.org/10.1016/j.mce.2020.110987

Cherouveim P, Velmahos C, Bormann CL (2023) Artificial intelligence for sperm selection-a systematic review. Fertil Steril 120(1):24–31. https://doi.org/10.1016/j.fertnstert.2023.05.157

Drevet JR, Hallak J, Nasr-Esfahani MH, Aitken RJ (2022) Reactive oxygen species and their consequences on the structure and function of mammalian spermatozoa. Antioxid Redox Signal 37(7–9):481–500. https://doi.org/10.1089/ars.2021.0235

Dutta S, Majzoub A, Agarwal A (2019) Oxidative stress and sperm function: a systematic review on evaluation and management. Arab J Urol 17(2):87–97. https://doi.org/10.1080/2090598X.2019.1599624

Fishman EL, Jo K, Nguyen QPH, Kong D, Royfman R, Cekic AR, Khanal S, Miller AL, Simerly C, Schatten G, Loncarek J, Mennella V, Avidor-Reiss T (2018) A novel atypical sperm centriole is functional during human fertilization. Nat Commun 9(1):2210. https://doi.org/10.1038/s41467-018-04678-8

Garanina AS, Alieva IB, Bragina EE, Blanchard E, Arbeille B, Guerif F, Uzbekova S, Uzbekov RE (2019) The centriolar adjunct⁻Appearance and disassembly in spermiogenesis and the potential impact on fertility. Cells 8(2):180. https://doi.org/10.3390/cells8020180

Lismer A, Kimmins S (2023) Emerging evidence that the mammalian sperm epigenome serves as a template for embryo development. Nat Commun 14(1):2142. https://doi.org/10.1038/s41467-023-37820-2

Manandhar G, Simerly C, Schatten G (2000) Centrosome reduction during mammalian spermiogenesis. Curr Top Dev Biol 49:343–363. https://doi.org/10.1016/s0070-2153(99)49017-9

Mannucci A, Argento FR, Fini E, Coccia ME, Taddei N, Becatti M, Fiorillo C (2022) The impact of oxidative stress in male infertility. Front Mol Biosci 8:799294. https://doi.org/10.3389/fmolb.2021.799294

Marinaro JA (2023) Sperm DNA fragmentation and its interaction with female factors. Fertil Steril 120(4):715–719. https://doi.org/10.1016/j.fertnstert.2023.06.001

Milne GL (2017) Classifying oxidative stress by F2-Isoprostane levels in human disease: the re-imagining of a biomarker. Redox Biol 12:897–898. https://doi.org/10.1016/j.redox.2017.04.028

Moretti E, Noto D, Guazzo R, Menchiari A, Belmonte G, Collodel G (2021) Centriolar defects, centrin 1 alterations, and FISH studies in human spermatozoa of a male partner of a couple that produces aneuploid embryos in natural and artificial fertilization. J Assist Reprod Genet 38(5):1197–1205. https://doi.org/10.1007/s10815-021-02109-0

Moretti E, Signorini C, Corsaro R, Giamalidi M, Collodel G (2023) Human sperm as an in vitro model to assess the efficacy of antioxidant supplements during sperm handling: a narrative review. Antioxidants (Basel) 12(5):1098. https://doi.org/10.3390/antiox12051098

Mortimer D (2018) The functional anatomy of the human spermatozoon: relating ultrastructure and function. Mol Hum Reprod 24(12):567–592. https://doi.org/10.1093/molehr/gay040

Noto D, Collodel G, Cerretani D, Signorini C, Gambera L, Menchiari A, Moretti E (2021) Protective effect of chlorogenic acid on human sperm: in vitro studies and frozen-thawed protocol. Antioxidants (Basel) 10(5):744. https://doi.org/10.3390/antiox10050744

Parra A, Padilla L, Lucas X, Rodriguez-Martinez H, Barranco I, Roca J (2023) Seminal extracellular vesicles and their involvement in male (in)fertility: a systematic review. Int J Mol Sci 24(5):4818. https://doi.org/10.3390/ijms24054818

Schlegel PN (2023) Can sperm quality affect reproductive outcomes? Fertil Steril 120(4):707–708. https://doi.org/10.1016/j.fertnstert.2023.08.007

Shi Z, Yu M, Guo T, Sui Y, Tian Z, Ni X, Chen X, Jiang M, Jiang J, Lu Y, Lin M (2024) MicroRNAs in spermatogenesis dysfunction and male infertility: clinical phenotypes, mechanisms and potential diagnostic biomarkers. Front Endocrinol (Lausanne) 15:1293368. https://doi.org/10.3389/fendo.2024.1293368

Vallet-Buisan M, Mecca R, Jones C, Coward K, Yeste M (2023) Contribution of semen to early embryo development: fertilization and beyond. Hum Reprod Update 29(4):395–433. https://doi.org/10.1093/humupd/dmad006

Vashisht A, Gahlay GK (2024) Understanding seminal plasma in male infertility: emerging markers and their implications. Andrology 12(5):1058–1077. https://doi.org/10.1111/andr.13563

MIX
Papier aus verantwortungsvollen Quellen
Paper from responsible sources
FSC® C105338

If you have any concerns about our products, you can contact us on
ProductSafety@springernature.com

In case Publisher is established outside the EU, the EU authorized representative is:
Springer Nature Customer Service Center GmbH
Europaplatz 3, 69115 Heidelberg, Germany

Printed by Libri Plureos GmbH
in Hamburg, Germany